Popper's Critical Rationalism

Routledge Studies in the Philosophy of Science

1. Cognition, Evolution and Rationality
A Cognitive Science for the Twenty-First Century
Edited by António Zilhão

2. Conceptual Systems
Harold I. Brown

3. Nancy Cartwright's Philosophy of Science
Edited by Stephan Hartmann, Carl Hoefer, and Luc Bovens

4. Fictions in Science
Philosophical Essays on Modeling and Idealization
Edited by Mauricio Suárez

5. Karl Popper's Philosophy of Science
Rationality without Foundations
Stefano Gattei

6. Emergence in Science and Philosophy
Edited by Antonella Corradini and Timothy O'Connor

7. Popper's Critical Rationalism
A Philosophical Investigation
Darrell P. Rowbottom

Popper's Critical Rationalism

A Philosophical Investigation

Darrell P. Rowbottom

Routledge
Taylor & Francis Group
NEW YORK AND LONDON

First published 2011
by Routledge

605 Third Avenue, New York, NY 10017
4 Park Square, Milton Park, Abingdon, Oxon OX14 4RN

Routledge is an imprint of the Taylor & Francis Group, an informa business

Typeset in Sabon by IBT Global.

First issued in paperback 2013

Library of Congress Cataloging in Publication Data

Rowbottom, Darrell P., 1975–
Popper's critical rationalism : a philosophical investigation / by Darrell P. Rowbottom.
p. cm. — (Routledge studies in the philosophy of science ; 7)
Includes bibliographical references and index.
1. Popper, Karl R. (Karl Raimund), 1902-1994. 2. Rationalism. I. Title.
B1649.P64R69 2011
149'.7—dc22
2010022916

ISBN13: 978-0-415-85006-3 (pbk)
ISBN13: 978-0-415-99244-2 (hbk)
ISBN13: 978-0-203-83618-7 (ebk)

I dedicate this book to my beloved little monster, Clara Aurelia, and to my long-suffering wife, Sarah.

Contents

Figures

Preface

An eminent colleague—one of the leading philosophers of science in the UK—once asked me, rhetorically, "What did Popper ever contribute to philosophy?" I was dumbstruck. I mumbled something about his axioms of probability, which certainly wasn't liable to be convincing, and quickly shrank back into the defensive posture with which I am, alas, familiar.

For the most part, Popper's work isn't taken all that seriously in contemporary Anglo-American philosophy (at least when one gets anywhere near the 'highest level', meaning the most prestigious journals). My personal experience (which is admittedly limited) is that it is easier to publish pieces which are critical, rather than supportive, of Popper's ideas. (As you will see, I have plenty to criticise as well as plenty to support.) And perhaps this is not such a surprise, given the way that his position is typically presented, again in my experience (and based on an impromptu survey of available lecture notes on the web), in undergraduate (and even postgraduate) lectures. He is painted as a simple falsificationist, following in the footsteps of the refuted logical positivists, and set up for a fall. Falsificationism doesn't work because of Duhem's thesis and probabilistic hypotheses. Corroboration is just a fig-leaf for induction. Game over. Enter Kuhn.

I have also noticed that sometimes Popper is not referred to where his work is obviously relevant, even by philosophers who know this work very well. There is one recent book like this, in particular, which is written by one of the great stars of philosophy of science. Feyerabend makes an appearance. Popper does not. Referring to Popper is simply not cool. (Referring to the work of some of his ex-students is also uncool. I was once advised not to quote David Miller in one of my papers. My colleague's fear was that this would make the piece harder to publish.)

There is no doubt an interesting story to tell about how this state of affairs has come about. I will speculate. Popper's influence was, for many decades, most impressive. And it is natural (and proper) for there to be a critical backlash in response to such dominance. But Popper was also, by many accounts, aggressive in his personal conduct in philosophical contexts (such as his infamous seminar) and also somewhat neurotic. (See, for instance, Agassi's autobiography.) He wound a lot of people up the wrong way; most notably he alienated Feyerabend, who devoted much of his philosophical career to attacking (and even mocking) Popper's work in one way or another. Of course, this sort of consideration *should not* have affected how Popper's work is viewed. Unfortunately, I believe it has. (Newton was exceptionally neurotic, but also exceptionally gifted.)

There is, however, a flip side to this. Many critical rationalists, perhaps responsively, seem to me to have become rather insular. Rarely, for instance, do they refer to work going on in mainstream epistemology (e.g. virtue epistemology or social epistemology) or contemporary philosophy of science (e.g. constructive empiricism and structural realism). (There are exceptions. These tend to be younger philosophers.) They do, of course, engage with those who directly criticise Popper's work. But most of the criticisms are old, and therefore the debates have become rather stagnant.

I believe that both groups are missing out. I think that contemporary philosophy has something to learn from Popper and that Popperians (or Neo-Popperians) have a lot to learn from contemporary philosophy. This book makes a start at bridging the gap, by developing critical rationalism in the light of more recent philosophical advances. Do not expect, then, a mere presentation of Popper's ideas. Expect a critical engagement with, and development of, those ideas. What I end up with is an articulation and defence of (part of) my own current philosophical position on science, which I see as an advancement of the (pan)critical rationalist research programme. I stand on Popper's shoulders, although I am sure that he would have vehemently opposed several of my proposals, most notably that anti-realism is the perfect bedfellow for anti-inductivism.

I have a lot of people to be grateful to for their help in improving this book. First and foremost is Peter Baumann, who commented on the whole book in draft. Second are those who have commented on chapters in draft: Gunnar Andersson, Alexander Bird, Tony Booth, Donald Gillies, Ward Jones, David Miller, John Preston, Duncan Pritchard, Paul Teller, and Tim Williamson. Third are those who have commented on the papers that form the basis of some of the chapters (and have not already been mentioned): Joseph Agassi, Otávio Bueno, Matthew Kopec, James McAllister, Bence Nanay, Elie Zahar, and numerous anonymous referees. Naturally, I also owe a debt of gratitude to those who supervised my work as a graduate student, namely Jonathan Lowe and Robin Hendry. (Amusingly, when I worked with the latter I was an ardent inductivist. Much of my master's thesis was devoted to attacking critical rationalism!) I should also mention Barry Gower, who played a significant part in inspiring my interest in scientific method.

On the publishing side, Erica Wetter has been a patient and supportive editor. And finally, I am extremely grateful to the British Academy for supporting my work by way of their Postdoctoral Fellowship scheme. If it were not for the research time that this Fellowship has afforded me, this book would not have appeared in its present form.

1 Comprehensive Rationalism, Critical Rationalism, and Pancritical Rationalism

The material in this chapter bears on one of the most important and fascinating problems in epistemology: is there any position, or belief set, that is not underpinned by faith (or that lacks dogmatic elements)? The worry is that if there is not, then relativism beckons; that it would be acceptable to choose whichever poison one likes (or to stick with whichever poison one inherits).

Almost all contemporary Western philosophers working in the analytic tradition, and plausibly most academics, subscribe to the view that there is something wrong about holding beliefs on the basis of faith alone. Thus the religious fundamentalist is taken to be blameworthy (from an epistemological perspective, at least) for uncritically accepting peculiar doctrines, or for making no attempt to subject her most deeply felt religious beliefs to scrutiny (with an eye to potentially renouncing them); that is, provided that she is not a victim of brainwashing.

But let's imagine we were to suggest to such a fundamentalist that she should criticise her beliefs by appeal to experience and her own faculty of reason, and she adopted a purely defensive strategy in response (i.e. tried merely to defend her entitlement not to do so). What would we say if she declared that she did not have *faith*, as we do, that using experience and reason enables one to determine the truth (or to achieve whichever other goal we happen to agree that we are after)? What if she intimated, that is to say, that we are simply possessors of a different faith than hers? She might even suggest that as God is her authority, so reason and experience are ours.

We might respond that she could only understand God's will by employing reason and experience, but this will not settle the matter. She may easily respond that she has faith in a particular class of experiences, whereas we have faith in a different class. Alternatively, she might suggest that we simply experience the world differently to her (in so far as experience is theory laden), or just that we have faith in a different interpretation of experience.

Even if she accepts that she has many of the same beliefs as us, derived from the same sources, e.g. when it comes to mundane matters, she may argue that revelation is an additional source of knowledge that operates in a peculiar realm where reason and/or experience are impotent. And so forth.

So she has plenty of room to manoeuvre. And her charge is just that when we reach rock-bottom, each of us can do nothing other than appeal to faith in order to defend our beliefs. As Wittgenstein (1963, §217) puts it:

> If I have exhausted the justifications, I have reached bedrock, and my spade is turned. Then I am inclined to say, "This is simply what I do."[1]

Can we resist this conclusion? If we cannot, then we must concede (at least if we maintain our faith in our reason) that we ought not to blame the religious fundamentalist for her beliefs. Furthermore, we should recognise that while we might try to persuade her of the wisdom of adopting our faith instead, employing argument is no nobler a path than using rhetoric, or perhaps even violence. For all our pretensions to epistemic and intellectual superiority, we would be zealots too.

I must now confess that I remain unconvinced that we *can* ultimately resist this conclusion. But (pan)critical rationalism is, as we will see, a noble attempt—and in my opinion, the best attempt—to do so. My goal in writing this book is partly to see if it succeeds. Before we come to this, however, we should first look at the competition.

1. COMPREHENSIVE RATIONALISM

In the second volume of *The Open Society and Its Enemies*, Popper (2003 [1945], p. 249) contrasted two forms of 'rationalism', each of which he understood to be 'an attitude that seeks to solve as many problems as possible by an appeal to reason, i.e. to clear thought and experience, rather than by an appeal to emotions and passions.'

The first, which he dubbed 'comprehensive rationalism', is possessed by the individual who will not 'accept anything that cannot be defended by means of argument or experience', and therefore implicitly obeys a rule such as 'any assumption which cannot be supported either by argument or by experience is to be discarded' (Popper 2003 [1945], p. 254). Popper found fault with this rule, however, because he judged that it cannot be coherently applied to itself. In short, he thought it was an assumption that cannot itself be supported either by argument or by experience, and which would therefore need to be discarded, paradoxically, if it were to be obeyed. Naturally one might instead suggest that the rule should be 'any assumption *except this one* which cannot be supported either by argument or by experience is to be discarded'. But then comprehensive rationalism would not meet its own exacting standards; quite to the contrary, it would appear to be rather dogmatic (as well as *ad hoc*). Popper therefore suggested a second form of rationalism, critical rationalism, to which we will come shortly.

Van Fraassen (2002) discusses a closely related problem. He examines the following thesis, which he calls Principle Zero:

> For each philosophical position X there exists a statement X+ such that to have (or take) position X is to believe (or decide to believe) that X+. (van Fraassen 2002, p. 41)

Van Fraassen then asks whether there is an E+ to the E of empiricism, while noting that it is a key component of empiricism that 'disagreement with any admissible factual hypothesis is admissible' (p. 43). Furthermore, he notes that E+ must provide the basis for empiricist critique:

> Empiricist critique of X = demonstration that X is incompatible with (contrary to) the empiricist dogma E+. (Ibid.)[2]

So if E+ is to be a factual hypothesis, which it seems it must be if empiricism is to be thoroughgoing, then disagreement with E+ is (rationally) admissible. Yet empiricist critique of E+ is impossible (because clearly E+ is compatible with E+). A comprehensive empiricism cannot meet its own exacting standards:

> Perhaps E+ . . . says something like "Experience is the only source of information," from which the empiricist may then derive "So there can be no a priori demonstration or refutation of any factual claim." But E+ is itself precisely such a factual hypothesis. This means that its contraries are also putative matters of fact. So they must be admissible in the same way as empirical hypotheses are generally in the sciences. Unfortunately, since E+ is the dogma that sums up the entire basis of this empiricism, it is also the sole basis for any empiricist critique . . . It follows now that by the empiricist's own lights, any empiricist critique can therefore be legitimately countered as follows: "The target of your critique is a claim contrary to E+, hence equally admissible as a hypothesis and not to be ruled out from the outset". (Ibid., pp. 45–46)

Let's now consider how Popper's and van Fraassen's discussions are connected. The fact that disagreement with any factual hypothesis is admissible follows from the reasonable assumption that information might tell either way. And what counts as (admissible) information is specified *by the factual hypothesis that constitutes the core dogma* of the position (such as "Experience is the only source of information"). Disagreement with the core dogma is admissible because it is factual (as interesting claims about what does or does not count as information always are).

So how do the rules discussed by Popper enter? The answer is that these are partially dependent on what counts as (admissible) information. (The form of the rule will be 'do not accept anything that cannot be defended by *x*' or the weaker 'do not accept anything contrary to *x*', or some such, where *x* is the source of information specified in the dogma.) But this means that disagreement with the rules is admissible, *as a result of the fact that disagreement with the core dogma is admissible.*

Van Fraassen's discussion of this issue has attracted considerable attention, and has been commented on by many contemporary philosophers.[3] But it is remarkable that its relation to Popper's argument against comprehensive rationalism has not been noticed.[4] It is best seen as a supplement, which indicates that something akin to the *tu quoque* discussed in section 3 applies even if rules for inquiry are taken to be applicable to the dogmas that underlie them (and from which they are derived). To pre-empt this discussion, it would seem that relativism beckons because each form of comprehensive rationalism based on a dogma is as defensible as any other.

2. RESPONSIBILITY FOR BELIEF

At this juncture, we should pause for a moment to consider what it means to be responsible for one's beliefs. It is plausibly true—indeed, this is a thesis called 'doxastic involuntarism' (see Alston 1989; Steup 2000, 2008; Jones 2003)—that we do not have direct control, in a wide range of circumstances at least, over what we believe. I cannot, for instance, choose to look out of the window in my study and fail to believe that I am looking at a reservoir. Nor can I elect to disbelieve momentarily that my pet rabbit is suffering from a serious illness (although doing so might help me to write rather more easily). However, it is plausibly true that we can intentionally put ourselves into situations where our beliefs are susceptible to change, or take actions which are liable to result in belief changes.[5] In order to form a belief about what the weather will be like today, I might either consult the barometer on my wall or visit the meteorological office's website. But if I were to form a belief in the former way rather than the latter, when each way was as easy as the other (and otherwise as desirable for utility-based reasons), then I may have made a mistake. So we may be responsible for what we believe (at least) with respect to whether we put ourselves into any potentially belief-altering scenarios at all. Someone who refused to put herself into a situation where her religious beliefs might be challenged, or avoided spending any time considering those beliefs, could therefore be blameworthy.

This is not, of course, quite as simple a matter as it might initially appear to be. Many of a religious fundamentalist's dogmatic beliefs about God might (happen to) be true. And perhaps, furthermore, she may strongly (and truly) believe that to subject herself to a situation where those beliefs would be challenged would lead her to a crisis of faith (and thereby to lose those true beliefs and gain false ones in their place). So from a decision-theoretic perspective, she would apparently be right not to question her religious beliefs. She would also have hit upon the truth anyway, so it would be good (even if only lucky) from the point of view of preserving true beliefs and eliminating false beliefs that she didn't question them. What this points to is a fundamental difficulty in explaining and isolating epistemic reasons for action—see Booth (2006, 2009) and Rowbottom (2008a)—and defending our intuition (which is central to the Socratic tradition) that being reflective is virtuous. For the moment, however, let's imagine that this intuition can be adequately defended; we will return to the matter towards the end of this chapter, where we will consider a positive argument for being (pan)critical.

3. CRITICAL RATIONALISM

Given the unsatisfactory nature of comprehensive rationalism, what is the rationalist—who seeks, recall, 'to solve as many problems as possible by an appeal . . . to clear thought and experience' (Popper 2003 [1945], p. 249)—to do? Here's where the second form of rationalism discussed by Popper, namely critical rationalism, comes in. In *The Open Society and Its Enemies*, he describes it as involving: 'fundamentally an attitude of admitting that "*I may be wrong and you may be right, and by an effort, we may get nearer to the truth*".' But the same sentiment is also clear in his earlier writing.[6] In the first English edition of *The Logic of Scientific Discovery* (in a passage which appears in the 1934 German original), one learns:

> A system such as classical mechanics may be 'scientific' to any degree you like; but those who uphold it dogmatically—believing, perhaps, that it is their business to defend such a successful system against criticism as long as it is not *conclusively disproved*—are adopting the very reverse of that critical attitude which in my view is the proper one for the scientist. (Popper 1959, p. 50)

A noteworthy feature of this quotation is that it refutes a naive 'falsificationist' interpretation of Popper's philosophy (which has not entirely vanished, sadly, when it comes to introductory university lectures).[7] In particular, it asserts that how we treat our theories is just as important as whether they are falsifiable in principle. So for Popper, one cannot be a genuine scientist by picking some falsifiable claim such as 'All rabbits are brown' and sticking to it come what may. (Nor can one be a genuine scientist by working only with unfalsifiable claims! One should bring one's theories into potential conflict with experience.[8])

But the worry now is that being a critical rationalist seems ultimately to be a matter of faith. In the following passage, indeed, Popper appears to admit as much:

> Whoever adopts the rationalist attitude does so because he has adopted, consciously or unconsciously, some proposal, or decision, or belief, or behaviour; an adoption which may be called 'irrational'. Whether this adoption is tentative or leads to a settled habit, we may describe it as an irrational *faith in reason*. (2003 [1945], p. 255)

If this is correct, however, then it hardly provides a platform on which to promote critical rationalism—*understood now as the philosophical position that we ought to adopt critical attitudes*—by argument. Instead, the way ahead would appear to be to proselytise; to spread the faith by encouraging people to make such an irrational leap. Evangelism and critical rationalism would go hand in hand, in principle if not in practice.

In fact, as was forcefully argued by Bartley (1962), the admission that we require '*faith in reason*' provides an excuse for irrationalism.[9] If one admits that being a rationalist requires an irrational move, that is to say, then one should also admit the primacy of irrationalism. There is hence a great excuse—by the rationalist's own admission—for selecting a different faith:

> In sum, the belief that rationality is ultimately limited, by providing an excuse for irrational commitment, enables a Protestant, or any other irrationalist, to make an irrational commitment without losing intellectual integrity. (Bartley 1962, p. 103)

Furthermore, it appears curious to suggest that a genuine rationalist would *want* to encourage anyone, rationalist or not, to perform a totally irrational act: to make a leap of faith or to subscribe to a dogma. "Behave irrationally just one more time, because it's the key to behaving rationally!" is no way to advocate rationalism. Nor, indeed, is "Be dogmatic in your non-dogmatism!"[10] Similarly, Christians who are true to the principles of Christianity do not resort to unethical means to win converts. (And if we are not prepared to make irrational leaps of faith, then why should we expect others to do so?)

Given the *tu quoque*, it is perhaps rather remarkable that passages such as the following are reasonably commonplace: 'The decision for empiricism as an act of scientific faith signifies that the best way to acquire reliable knowledge is the way of evidence obtained by direct experience' (Barratt 1971); and 'scientists properly have a sufficient degree of faith in their basic theoretical postulates . . . that anomalies are explained away' (Newton-Smith 1981, p. 81).[11] If science rests on faith, then why should we prefer it to astrology?

3.1 Van Fraassen's Alternative to (Dogma-Based) Comprehensive Rationalism

Before we continue by seeing how Bartley attempted to avoid the *tu quoque*, perhaps we should see if we can learn anything from van Fraassen's response to the problem with comprehensive rationalism (and comprehensive empiricism in particular) when this is construed as resting on a dogma (or set of possible dogmas). His radical alternative is to reject the notion that philosophical positions should be understood as resting on claims, or collections of claims. Instead, he proposes that we understand them as 'stances', which are not 'identifiable through the beliefs involved, and can persist through changes of belief' (van Fraassen 2002, p. 62). More particularly, 'other than factual theses', he suggests that such stances involve 'attitudes, values, commitments, [and] goals' (ibid., p. 48).

Van Fraassen (2004b) agrees with the thrust of Teller's (2004) idea that a stance may be understood as an *implicit* epistemic policy, and approves of my recent attempt with Otávio Bueno—see Rowbottom and Bueno (In Press A) and van Fraassen (In Press)—to explicate stances in terms of *modes of engagement* and *styles of reasoning* as well as propositional attitudes. Broadly, a mode of engagement is a way of approaching things; one may be active and 'hands on' like Faraday, or rather more passive and contemplative like Einstein, for instance. A style of reasoning involves (implicit or explicit) use of peculiar inferential strategies, or templates for problem/puzzle solving (similar to Kuhnian exemplars).[12]

So if a stance such as empiricism can be identified in terms of modes of engagement and styles of reasoning, rather than via beliefs (or other propositional attitudes), then one can be an empiricist without being a comprehensive rationalist *at least in so far as one need not commit to belief in any particular proposition, come what may.*[13] It is not quite as clear, however, that one may be an empiricist while failing to *implicitly* obey a rule such as 'any assumption which cannot be supported either by argument or by experience is to be discarded' (Popper 2003 [1945], p. 254). (Or a highly similar rule, such as 'Only assumptions that cannot be presently dismissed by argument or experience are to be entertained'.[14]) In short, it would seem that some forms of comprehensive rationalism may be construed as stances, although comprehensive rationalists *are typically committed to a dogma in virtue of which they obey the aforementioned rules.* This is essentially the worry I expressed in Rowbottom (2005), although I was especially concerned with whether van Fraassen's personal empiricism is a form of comprehensive rationalism.

However, we are here concerned not with empiricism as such, but rather with whether a critical position is tenable without faith, i.e. in such a way as to avoid a *tu quoque* objection. And on this issue, since positions are construed as being something other than propositional, there is the potential for a real breakthrough.

Now the critical attitude might be thought of primarily as a mode of engagement (which may preclude certain styles of reasoning). So any attempt to launch a *tu quoque* against the possessor of such an attitude would work only if said attitude were itself maintained on the basis of faith, i.e. in such a way that it could not be renounced on the basis of the activities and/or actions that it was responsible for (such as careful reflection on the rule, made explicit, that 'One should hold everything open to criticism'). The critic's point would be not that we all have to have faith in (some of) our beliefs, but rather that we all have to have faith in our own stances (or identifying components thereof, i.e. components other than propositional attitudes).[15]

Van Fraassen (2004b, p. 191, fn. 14) suggests not only that 'one is generally less committed to a policy than a stance', but also that:

> the term "stance" has its own connotations of commitment and intention: specifically, *the commitment to preserve oneself in that very stance* . . . There is a pragmatic inconsistency in "I am committed to doing x but not committed to maintaining this commitment."

If this were right, then it would mean that the possessor of a critical stance would be committed to maintaining that critical stance. But I do not think that this is correct, at least in so far as such commitment need stretch to dogmatism, as the following example illustrates.[16]

If I am committed to meeting you for dinner next week, but you phone me to cancel because one of your relatives has just passed away, then it would be wrong to suggest that I ought to be committed to maintaining the original commitment to meet you for dinner, on pain of inconsistency; that I ought to urge you nevertheless to meet me for dinner, so as to allow me to maintain my commitment to meet you for dinner. On the contrary, there would seem to be no inconsistency, pragmatic or otherwise, in simply discarding the commitment. And if you phone me back an hour later in order to ask me whether I would still meet you for dinner only to be informed that I now have different plans, you ought not to accuse me of inconsistency!

In response, one might insist that the commitment to preserve the commitment (or the second-order commitment) simply vanishes at the moment the (first-order) commitment vanishes, i.e. when you cancel. But imagine instead that I decide that I would prefer not to go to dinner with you, despite our prior arrangement, because I'm feeling rather tired. Would I be wrong (in any sense whatsoever) to ask you to release me from the commitment to meet you for dinner (while nonetheless being committed to fulfilling my obligation to meet you, while it remains in existence)? Surely not, although this would involve actively seeking to remove, rather than preserve, the commitment! It therefore seems that one may be committed to doing x while being simultaneously committed to losing the commitment to do x (in at least one possible way).

Van Fraassen might rejoin by giving a different spin on the example, say, on the question of whether one would allow oneself to be brainwashed, given the understanding that this would 'remove' the commitment. But this is rather different: to do one's best to fulfil one's commitment (which could perhaps remain even if one were unaware of it) is not to do one's best to *preserve* that commitment. In fact, fulfilling one's commitment often results in the termination of that commitment (so at best, van Fraassen should have written "I am committed to doing x until I have done x but not committed to maintaining this commitment"). And perhaps fulfilling a commitment to the critical attitude actually *entails* abandoning—i.e. being willing to fail to preserve—that attitude under appropriate circumstances. In fact, I think that van Fraassen should be delighted to concede this point, because the notion of positions as stances then provides resources for solving a classic problem in a novel way. (One might object that only attitudes, and not stances, are required. But whereas Popper insisted that critical rationalism is an attitude *rather than* a position, van Fraassen's work suggests that *a philosophical position can be construed as centrally involving an attitude*.)

However, there is still no argument that one *ought to adopt* a critical stance. As it happens, van Fraassen allows that very many different stances are permissible. (So he may seek to persuade us to adopt the empirical stance, but will not accuse us of being 'irrational', or in some sense epistemically blameworthy, if we do not.) But this view, which is often referred to as 'stance voluntarism', has come under attack—e.g. by Chakravartty (2004, In Press), Rowbottom (2005), and Baumann (In Press)—precisely because it appears to open the floodgates to irrationalism. A limit on permissible stances, one which gives considerable room for movement (and allows, for instance, that both empiricists and some of their rivals may be rational) while simultaneously forbidding dogmatism, would appear to be required.

One might therefore suggest that the critical attitude is a crucial part of any epistemically permissible stance (and that we are responsible for the stances we hold because we have some control over this); in fact, Rowbottom and Bueno (In Press A) argue precisely this. But as we will see later in this chapter, there is a more direct argument, which makes appeal only to the notion of reliable belief-forming processes, that we should adopt the critical attitude in an important class of contexts (e.g. when in the process of inquiry). For the moment, let us return to the issue of critical rationalism.

4. THE LIMITS OF COMPREHENSIVE AND CRITICAL RATIONALISMS: PANCRITICAL RATIONALISM TO THE RESCUE?

A fundamental problem in classical epistemology is to find an appropriate terminus to resist persistent requests for justification (such as "Why believe in what your senses make you think?"), or to elaborate and defend an account of *basic knowledge* (in modern parlance).[17] But all of the obvious strategies have obvious defects. We could say that some statements are self-evidently true, or self-justifying. However, this would appear to deny us resources to deal with scenarios where we disagree on which statements fit the bill. If I boldly assert that it is self-evident that the law of non-contradiction holds, but Graham Priest asserts precisely the opposite, then what room for manoeuvre do we have? You might think that we could have ultimate recourse to the propositions which we both agree are self-justifying. But what if there were none of those (above and beyond the statement, perhaps itself supposed to be self-justifying, that some statements are self-justifying)? Like two religious fundamentalists with opposed and irreconcilable beliefs, we might find ourselves at loggerheads for the remainder of our days. There would be no obvious way for either of us to make rational progress in collective inquiry, or individual inquiry in so far as that should involve taking seriously the possibility that one's opponent might be correct.[18] (Of course, this doesn't show that there aren't self-justifying propositions. It's just that even if there are, it won't help us in the business of inquiry. If I claim that I can infallibly identify these propositions and that you cannot, but you claim the opposite, then how should we adjudicate?[19])

Another well-known strategy is to suggest that the coherence of some set of propositions serves to justify them; that the greater the coherence of one's set of beliefs, the greater justification one has in holding it (and each of its members).[20] But the well-known problem with such a strategy is that there appear to be many, plausibly even infinitely many, maximally consistent sets of possible beliefs. Thus it would be possible to select some small consistent set of beliefs as one's starting point and then erect a superstructure of coherent beliefs. One could then allow the superstructure to change while refusing to abandon the set of dearly held beliefs. (This is rather like the lesson we learn from Duhem's thesis; we can defend our pet beliefs/theories by abandoning/altering our auxiliary beliefs/hypotheses.) Of course, the fan of the coherence account may counter with the suggestion that sometimes beliefs are 'forced upon us', e.g. by acts of perception. But then a *pure* coherentist account is being abandoned, and the remainder of the discussion here, either concerning self-evidence (preceding) or externalism (following), becomes pertinent.

So remaining on the topic of perception, why not say that some statements count as items of basic (and perhaps even unassailable) knowledge because of the unintentional processes by which they are classified as true? Why not say that if my true belief that there is a table in front of me is generated by an appropriate causal process, say, then it counts as basic knowledge? *The answer is not so much that we should not say this, but rather that doing so does not appear to solve the present problem satisfactorily.* Imagine for the sake of argument that there really is such basic knowledge; e.g. that most of our reports about our own sensory experiences are true. If we are concerned with the process of inquiry, and how we should proceed, we will wish to identify basic knowledge (and its sources). (It would be useful in science to have a stock of basic statements that were true. These could serve as potential falsifiers.) But now we need to ask how we are supposed to do this. If we assume that whenever we know something then we are aware that we know it, there won't be any problem. But why should we assume this?[21] If we have to argue for it, then will the argument itself rest on items of knowledge? And even if it does, won't *whether* it does be the subject of further questioning?[22] In short, the objection here is the classic one that the internalist (who thinks that if there is to be any form of justification, we must have access to it) makes against externalism. Imagine we miraculously knew beyond all doubt that either science or Bible studies (but not both) provided a reliable means by which to generate true beliefs (and eliminate false beliefs). Which should we choose? Externalism, at least when understood in the relatively crude fashion outlined earlier, appears to be silent.[23] And in our actual lives, there are very many possible sources of knowledge (or processes by which to form/change beliefs). I could consult tabloid newspapers, read tea-leaves, employ tarot cards, and so on. If whether we are justified or not is independent of our personal conduct, then it is difficult to see how we can be held blameworthy for our conduct in inquiry. A dogmatist who believed only in those things that he had experienced with his own senses would be better off, if perception is a reliable belief-forming process, than a Socratic philosopher who pondered questions concerning the world beyond the observed and sometimes doubted that his experiences disclosed the way that things really were.

This pattern of thought led Bartley to conclude that justificationism—broadly, the view that beliefs should always be justified in order for them to be reasonable—is a fundamental philosophical error. To summarise, the problem is that it always appears to be OK, according to such a philosophical viewpoint, to keep on asking "Why?" And the only way to go when you follow this strategy through is either back forever, into an infinite regress, or around forever in a circle. The way out of such a difficulty, as we have seen, is just to choose a place to stop (or some circle to stick with). But such a choice would appear to be arbitrary, and therefore dogmatic, according to the very doctrine of justificationism. (Nonetheless, it is worth noting that neither Bartley nor Popper gave externalism due consideration in print. Presumably they were already thoroughly disenchanted with the entire debate in mainstream epistemology by the time that work such as Goldman [1967], Armstrong [1973], and Goldman [1975, 1979] appeared. In the final part of this chapter, I will try to rectify this oversight by showing how externalism can play a part in pancritical rationalism.) In the words of Bartley (1984, p. 123):

> The classical problem of rationality lay in the fact that, for logical reasons, the attempt to justify everything (or to criticize everything through logical justification) led to infinite regress or dogmatism. But nothing *in logic* prevents us from holding everything open to nonjustificational criticism. To do so does not, for instance, lead to infinite regress.

The key idea is that it is possible to hold all one's positions open to criticism, *including one's position that one can and should hold all positions open to criticism*, in principle. (And naturally the absence of justification does not count as criticism; in short, criticism and justification are decoupled.) In this way—by being *pancritical*—one can avoid being susceptible to the *tu quoque* argument.[24]

To summarise, the problem with comprehensive rationalism is the presumption of justificationism. The problem with critical rationalism, as espoused by Popper, is the admission that trusting in experience and reason is irrational *because doing so is unjustified.* (Critical rationalism is only better than comprehensive rationalism in so far as it is humbler and not self-contradictory.) Moving criticism to centre stage and decoupling criticism and justification—as Popper had already done in his treatment of science, long before Bartley pointed out that this could extend to epistemology more generally—is the solution.

For most (analytic) philosophers, it is still extremely difficult to understand (let alone to sympathise with) this radical move. (And even those who advocate it sometimes find themselves thinking, against their best intentions, in a justificationist fashion.) The (understandable) knee-jerk reaction is that it opens the floodgates to irrationalism. The pancritical rationalist's response, however, is that the situation is quite the opposite. Justificationism opens those gates—in a way that gives an illusion of security—whereas the choice to abandon it provides us with an opportunity to slam them shut. But with due humility, the pancritical rationalist will confess that it may be impossible to shut them altogether; and in that event, we're all in the same boat!

5. CRITICISMS OF PANCRITICAL RATIONALISM

Having seen the rationale behind pancritical rationalism, we should now consider how it stands up to criticism itself. Unsurprisingly, it has been the subject of many discussions by a variety of philosophers, including, but not limited to, Watkins (1969, 1971), Kekes (1971), Richmond (1971), Agassi et al. (1971), Post (1972, 1983), Bartley (1983; 1984, app. 4), Helm (1987), Hauptli (1991), Miller (1994, §4.3), and Rowbottom (2009b). In what follows, we will therefore only be able to focus on the key criticisms as I see them.[25]

5.1 Watkins's Irrefutability Objection

Putting the previous discussion of stances to one side, let's assume for the moment that pancritical rationalism may be summed up neatly by a statement such as 'A rationalist can and should hold all his positions open to criticism' (Watkins 1969, p. 58). The question now arises of how, precisely, it is possible to hold that statement of pancritical rationalism (since it is self-referential) open to criticism. According to Bartley (1984, p. 120):

> [S]omeone could devastatingly refute this kind of rationalism if he were to produce an argument showing that at least some of the unjustified and unjustifiable critical standards necessarily used by a pancritical rationalist were uncriticizable to boot, that here, too, *something* had to be accepted as uncriticizable in order to avoid circular argument and infinite regress.

So in short, one might refute pancritical rationalism by indicating that some of its components are uncriticisable. Watkins's (1969, p. 57) response to this, however, is to suggest that pancritical rationalism therefore 'turns out . . . to be a perfect example of . . . a *dictatorial strategy* . . . [because] a defender . . . can be assured of victory over his critics however good their criticisms may be'. In short, Watkins's basic idea is that one can only show that pancritical rationalism is uncriticisable by criticising it:

> To put it in one sentence: in support of the claim that CCR [i.e. pancritical rationalism] is criticisable we are challenged to *criticise* it in a certain way—namely, by trying to show that it is *uncriticisable*! (Watkins 1969, p. 60)

As Bartley (1984, pp. 241–246) makes clear in his response, however, this argument misses the target. First and foremost, the mere existence of a means by which to deflect criticism does not show that one should, let alone that one must, employ said means. So someone claiming to be a pancritical rationalist *could* deflect criticism of her position by declaring "Every attempt to show my position is uncriticisable only criticises it, and therefore confirms it!" But this hardly means that she *should* do this; and it would be absurd to suggest that she must (even if only on pain of violating some epistemic norm). If she did, this would play into the critic's hands. Not only would she have access to a criticism deflecting measure, but she would also be actively employing it.

As an analogy, consider a scientist who is faced with experimental results that appear to refute his pet theory, which he has spent his entire life working on. Rather than accept that the results refute his theory, he may instead declare that the auxiliary assumptions used in testing it are false. So rather than accept that Newton's laws are falsified by the anomalous orbit of Mercury, say, he might continue to posit *ad hoc* explanations (e.g. the existence of unknown, invisible yet massive, heavenly bodies) to the end of his days. But does this show that it is false to say that "one can and should hold any theory open to criticism"? Surely not! What it shows, rather, is that there is a particular criticism-deflecting strategy that *can* be employed in order to shield general empirical hypotheses (in the form of universal statements, perhaps) from refutation.[26] As Bartley (1984, p. 242) puts it:

> If someone were to come forward with a cogent argument against pancritical rationalism; and if I were then to reply: "Oh, you see, that just goes to show that I was right in saying that my position is open to criticism", I would be laughed at. And we all know from Charlie Chaplin that the one thing that dictators cannot stand is to be laughed at . . .

Second, moreover, a critic's point can stand irrespective of whether the self-styled pancritical rationalist uses a peculiar measure to deflect criticism. And although the fact that a point stood *would* indicate that pancritical rationalism was criticisable in some way(s), it would not indicate that pancritical rationalism was criticisable in all salient ways. As Bartley (1984, p. 243) explains:

> A system that is uncriticizable is uncriticizable in some *particular, specific*, respects. That is, it must use a *particular* criticism-deflecting stratagem; it must use a *particular ad hoc* device, and so on. And the critic will, of course, identify these. Although the fact of this particular criticism will show that the system is in *some* respect criticizable, it won't make those of its features which diminish its criticizability, or which even render it virtually uncriticizable in particular circumstances, disappear.

A similar point was also made earlier, by Agassi, Jarvie and Settle (1971, p. 44):

> An admissible proof that CCR [read 'pancritical rationalism'] was in one particular respect uncriticisable would be proof that in some other respect CCR was criticisable. But this is innocuous: not even thoroughgoing criticisability is a sufficient condition for holding CCR; on the contrary, particular uncriticisability may be sufficient to force reasonable men to reject it.

In summary, to show that pancritical rationalism is uncriticisable in one respect is to criticise it in another respect. But showing it is uncriticisable in that first respect may suffice to convince one of its proponents to abandon it.[27]

We have seen that pancritical rationalism cannot fairly be called a 'dictatorial strategy'. In passing, it is worth mentioning that Watkins (1969) also offered examples of allegedly uncriticisable statements. As Bartley (1984, p. 244) points out, however, even if there *are* such statements it does not follow that one must rely on them in order to criticise (or indeed to be a pancritical rationalist).[28] Furthermore, some of Watkins's examples rely on the assumption that a firm distinction between the analytic and the synthetic can be maintained; but following Quine (1951), this presumption is dubious.

5.2 Post's Semantic Paradox

Post (1972, 1983) suggested that pancritical rationalism succumbs to something similar to the paradox of the liar. The resulting discussion proved rather long (and even torturous), however, so it will prove easier just to consider a version of Post's objection which Bartley (1984, p. 224) suggested was simpler but just as effective (and does not involve any assumptions that might be undesirable for the pancritical rationalist).

Consider the following two propositions:

(A) All positions are open to criticism.
(B) (A) is open to criticism.

If (B) were shown to be false then (A) would be shown to be false, because (B) is a consequence of (A). So we would have (successfully) criticised (A). But if (A) had been (successfully) criticised then it would have been shown to be open to criticism! So (B) would be true after all. In the words of Bartley (1984, p. 224): 'Any attempt to criticize (B) demonstrates (B); thus (B) is uncriticizable, and (A) is false'.

Bartley's response was not very satisfactory, in so far as he simply suggested—after pointing out that he nevertheless had no faith in (B) (and indeed didn't '*like*' or '*believe in*' it)—that this was an instance of a general problem with self-referential statements which was waiting to be satisfactorily resolved. In his own words:

> *The mere possibility of such a solution to the semantical paradoxes makes (B) criticizable after all: it suggests a potential means for invalidating the argument that produces the conclusion that (B) is uncriticizable.* And thus *Post* is refuted! (Ibid.)

In fact, this is an uncharacteristically weak argument from Bartley. Almost all of us would admit the *epistemic* possibility of a solution to semantic paradoxes. But it is possible to do so without also admitting that there *really is a solution out there waiting to be found* (i.e. that the solution is possible in a non-epistemic sense). And to admit the epistemic possibility of a solution is only to admit the *epistemic possibility* of invalidating the argument that produces the conclusion that (B) is uncriticisable. It is not to say "It can almost certainly be invalidated . . . we're just waiting to find out how!" Besides, Bartley (1984, p. 122) elsewhere suggested precisely that arguments should be *produced* (not merely expected) before they are considered:

> One will not begin to question statements that seem to be true simply in the face of arguments that it is, say, *logically* possible that they are not! In that sense, one calls a halt to criticism. One will, however, begin to question this "halting place" when a *particular* argument is produced to challenge it—when an argument is produced that renders it *problematical.*

However, Miller (1994, p. 89) suggests an alternative strategy for countering Post's objection. He points out that (A) only concerns *positions*, but that (A) may not itself be a position. And in that event, (B) does not follow from (A). (In fact, this is made even clearer if we consider positions to be stances, following van Fraassen but contrary to Miller, as earlier suggested.) By way of contrast, he invites us to consider:

(M1) All statements are open to criticism.
(M2) M1 is open to criticism.

In this case, Miller suggests that we should conclude that M1 is false (and that M2 is uncriticisable). He continues, however, by suggesting that this presents no problem for the pancritical rationalist:

> As far as statements . . . are concerned, what is important for the rationalist, I suggest, is that each statement that he accepts either is itself criticizable or follows from a statement that he accepts that is criticizable. Any position adopted must be criticizable, but it is no concession to the irrationalist to allow that some logical consequences of the position may not be criticizable. (Ibid.)

To this we may add that if positions are indeed something like stances, i.e. not reducible to (or dependent for their identity on) propositional attitudes, then they have no *propositional* consequences (i.e. entail nothing) whatsoever. But this does not render them uncriticisable. An implicit strategy may be rendered into an explicit policy, and its suitability for achieving particular aims may be examined.

It is worth adding that Bartley (1984, p. 223) also offered a general rejoinder to both Watkins and Post, which is helpful in order to summarise what we have covered so far in this section:

> [W]hen I declare that all statements are criticizable, I mean that it is not necessary, in criticism, in order to avoid infinite regress, to declare a dogma that cannot be criticized (since it is unjustifiable); I mean that it is not necessary to mark off a special class of statements, the justifiers, which *do* the justifying and criticizing but are not open to criticism; I mean that there is not some point in every argument which is exempted from criticism; I mean that the criticizers—the statements in terms of which criticism is conducted—are themselves open to review.

5.3 Sources of Criticism (Helm)

My own greatest concern about the viability of pancritical rationalism is rather different from those already covered, and far more pressing. As a way into seeing it, consider that in advancing criticisms we need to employ particular sources, e.g. personal observation reports based on (theory-laden) sensory experience, or formal systems (such as classical logic).[29] But we must be careful not to be dogmatically committed to using those sources or systems, if we are genuinely to be pancritical rationalists. Indeed Bartley (1984, p. 116) saw this too:

> Another strategy of criticism which is quite popular . . . also fuses justification and criticism . . . What matters is not whether a belief can be derived from the rational authority but whether it *conflicts* with it. In other words, it is not irrational to hold a belief that cannot be derived from—i.e., justified by—the rational authority unless its denial *can* be derived from the rational authority.

However, now consider a situation where we disagree about which sources of criticism are appropriate. If our disagreement is minor, and we agree on *some* sources (and systems), then perhaps we can use those to settle our dispute about the others on which we disagree. (Imagine one of us believes in the critical value of intuitions and the other doesn't, but we both trust in observation statements acquired by sensory experience. We can then use the latter, plausibly, to criticise one another's views on intuitions.) But if our disagreement is major, and we disagree on all sources (and possibly all systems) then there would appear to be an impasse. Just as those who believe in self-justifying propositions may disagree on what these are, as we saw earlier, so those who believe in proceeding critically may disagree on how to do so. A similar point is made by Helm (1987, p. 27):

> Bartley is not comparing like with like. He says that the rationalist can consider and be moved by criticism of logic and rationalism. But suppose that someone said that logic was defective because it was at odds with the word of God. Would the critical rationalist be moved by such a criticism? Presumably not. Again, Bartley says that the Christian theologian cannot, from his own Christian point of view, consider and be moved by criticisms of his Christian commitment. But why not? . . . why could not the Christian theologian be moved by criticism which stems from the word of God? Provided that like is compared with like, the two cases are fully symmetrical, and the *tu quoque* argument remains.

The resultant worry, which I take to be the most serious threat to pancritical rationalism, is along the following lines. Do we not, therefore, have license to freely choose which sources we accept? And why should that choice not be a matter of faith?

First, I think we should recognise that the prospects for pancritical rationalism—in the quest to avoid reliance on faith—are still considerably better than those for comprehensive rationalism and critical rationalism. Just because one starts with some set of arbitrary sources does not mean that one will stick with them come what may, and in that limited sense, at least, no faith in those sources would appear to be required. So even if we accept that the different parties in Helm's example have marked off their forms of allowable criticism in a way that allows them to avoid engaging with one another at a peculiar point in time, Helm does not show that those different parties *have a legitimate excuse for disagreeing, come what may, on which forms of criticism are allowable*; and he does not, therefore, succeed in showing that a *tu quoque* argument can still be run by the irrationalist.[30] A genuine pancritical rationalist will not dogmatically hold that there are only certain forms of allowable criticism; on the contrary, she will be willing to have her mind changed about what counts as a good (form or source of) criticism and what does not.

Second, and moreover, an individual with a critical attitude can attack someone else's views *from within the other's system (or by using the other's sources)*. One may attack empiricism as an all-encompassing philosophical system, that is to say, while relying on the same tools used by empiricists (such as classical logic and experience); and much critique may take such an immanent format. One may also endeavour to step inside the other's system and understand how one's own views look from that perspective. In the end, one's explorations might lead one to convert—or to adopt an entirely new system. No faith is required. (And Christians may have critical attitudes too, and change their own positions, of course. Many do.)

6. WHY ADVOCATE PANCRITICAL RATIONALISM?

We have seen that pancritical rationalism appears to stand in the face of the criticisms considered earlier. But simply because there is therefore no excuse for irrational commitment, it does not follow that we *should* be pancritical rationalists.[31] In fact, Bartley (1962, pp. 215–216) confesses: '[M]y argument does not force anyone to be a rationalist . . . Anyone who wishes, or who is personally able to do so, may remain an irrationalist.'

In saying that he has no argument against being a dogmatist, Bartley is perhaps a little unkind to himself.[32] In fact, it is possible to reconstruct two separate arguments from *The Retreat to Commitment*. The first of these is ethical in character, and suggests that those with critical attitudes may conduct themselves in a more understanding and generous manner than those without:

> [S]ince the rationalist . . . need be committed neither to his rationalism nor to any other of his beliefs, he need not repudiate people with whom he fundamentally disagrees. In principle, he can act toward them in a remarkable way. (Bartley 1962, p. 216)

The second argument is epistemic in character, and suggests that there are reasons of self-interest for having a critical attitude. The fundamental idea is that we may profit from our interactions with others in a manner that others may not from their interactions with us:

> [I]f we treat our opponents in discussion *not as they treat us, but as we would have them treat us,* it is we who profit . . . We may learn from the criticisms of our opponents even when their own practice prevents them from learning from us. (Ibid., p. 220)

There is a clear sense in which these two arguments are related. By adopting a critical attitude, we can behave in a way that enables us to learn from others when we otherwise might not be able to. And behaving in that way will also allow us to be gentler and kinder than we otherwise might be. We need not treat opponents as enemies of the faith, to be converted or dispensed with. We might even feel rather sorry for some of them (although not, of course, on the basis of a smug assumption of superiority on our own part).

In what follows, I will focus on the second argument; roughly, that there are situations in which we can learn if we're critical, but cannot if we're not. If correct, however, the first argument is still important. For it suggests that increasing the number of people with critical attitudes need not have a detrimental effect at the societal level; that one person's gain in this respect need not result in another's, or indeed the community's, loss.

Before we continue, however, I'd like to cast some doubt on the strength of the second argument, as it stands. Let's accept that having a critical attitude enables new learning possibilities; e.g. that a fundamentalist's belief in 'God exists' may not be shaken by any argument, whereas a pancritical rationalist's might be. What, precisely, is the benefit for the latter? 'God exists' may, after all, be true. And if we accept that we can learn things which are false—as I once learned, at school, that (well-educated) Europeans in the time of Columbus thought that the Earth was flat—then there may be no advantage whatsoever (or even a serious disadvantage in some scenarios).[33] In short, 'learning' in this (non-factive) sense—simply changing one's beliefs—doesn't seem to have any epistemic value.

The previous point may be made more starkly as follows. For every conceivable person with a large number of false beliefs and a small number of true beliefs who will come to have more true beliefs than false ones through possessing a critical attitude, there is a conceivable person with a large number of true beliefs and a small number of false beliefs who will come to have more false beliefs than true beliefs. Ranging over possible dogmatists, what's more, some lucky souls will be dogmatic just about those things that are true. (Perhaps one might appeal to the virtue epistemological notion, that 'success through virtue is more valuable than success by accident' [Greco 2008], but we will come to this in the next subsection. For the moment, suffice it to say that it is not clear how being critical *is* being virtuous, in so far as the end of minimizing false beliefs and maximizing true beliefs is concerned.)

In light of this, one option would be to suggest that having a critical attitude provides a kind of internal justification for one's beliefs that having an uncritical attitude does not. But even if this were true, what would the value of such justification be? BonJour (1985, pp. 7–8) suggests that there is none unless justification has a link to truth:

> Why should we . . . *care* whether our beliefs are epistemically justified? . . . The goal of our distinctively cognitive endeavors is *truth* . . . If finding epistemically justified beliefs did not substantially increase the likelihood of finding true ones, then epistemic justification would be irrelevant to our main cognitive goal and of dubious worth . . . Epistemic justification is therefore in the final analysis only an instrumental value, not an intrinsic one.[34]

Such a link between having a critical attitude and being in a better position to find the truth is, in fact, precisely what is missing. Caring about the truth and doing the best to find it might even lead you unerringly to believe things that are entirely false. Conversely, not caring a jot about the truth might lead you to commit dogmatically to a whole host of true beliefs.

Similar complaints—that there is a gap between method and aim—have been made about pancritical rationalism, and indeed Popper's philosophy of science more generally, before. Newton-Smith (1981, p. 70), for instance, attacks the putative link between corroboration and verisimilitude (or passing tests and being truthlike; we will return to this issue in the next chapter). And Watkins (1997, §13) suggests, along similar lines, 'If one is to aim at X, and pursue one's aim rationally, one needs to be able to monitor the success or failure of one's attempts to achieve X', and therefore proposed a negative answer to the question 'Are Popperians entitled to claim that one could do so if X were simply *truth*?'[35] Watkins thinks that the answer lies in appeal to *possible truth* as an aim. But surely this could also be the aim of dogmatic commitment, if fallibilism is accepted. It is, after all, more than merely *logically* possible for theories to be falsified on the basis of false observation statements. It is also more than *logically* possible to be committed (arbitrarily) to the truth of a proposition which is actually true.

Miller's (1994, p. 418) reply appears to be that 'falsificationism is unable to justify (in whole or in part) its role in the search for truth', and he would presumably add that this goes equally for the critical attitude. But this is hardly a satisfactory response. We are asking precisely whether there is a link between adopting a critical attitude and improving one's epistemic lot in some sense (even if this simply involves avoiding some errors). And it would appear to be perfectly reasonable to think that there is not such a link, if a suitable possibility cannot even be outlined. (If preferred, we are asking the pancritical rationalist to provide an argument against dogmatism. We have seen that "Dogmatists cannot learn" does not appear, taken alone, to do the job.)

As a way into finding an answer, note that to advocate pancritical rationalism is to reject authoritarianism, and indeed authoritarian accounts of justification, but need not be to reject the notion of justification wholesale. Consider, in this regard, Bartley's account of 'justificationism':

> [I]t is the view that the way to criticize an idea is to see whether and how it can be justified . . . Such justification involves:
>
> (1) an authority (or authorities), or authoritatively good trait, in terms of which final evaluation is to be made;
> (2) the idea that goodness or badness of any idea or policy is to be determined by reducing it to . . . the authority (or authorities), or to statements possessing the authoritatively good trait. That which can be so reduced is justified; that which cannot is to be rejected. (1984, pp. 186–187)

In fact, it is easy to see that one could fail to be a justificationist, in exactly this sense, while still believing in justification.[36] One need only believe that there are no authorities, or authoritatively good traits, according to which final evaluations of hypotheses—or even everyday statements—should be made. In short, one may accept the possibility that one's means of evaluation are not beyond question, and are in no sense 'final', while nevertheless accepting that justification is to be had. And one obvious way to do this, although surely not the only one, would be to suggest that one can be justified in believing that *p without realizing that one is justified in believing that p*.

But how does this help? Consider the notion that there are reliable means by which to form beliefs (or classify statements as true or false) which serve to (externally) justify the beliefs so formed.[37] Imagine, for instance, that we can employ a procedure which has a high propensity to accurately sort a peculiar class of propositions into 'true' and 'false' groups. A dogmatist will not be able to accept the results of such a procedure if they conflict with his commitments. Moreover, a dogmatist committed to a procedure that instead has a high propensity to inaccurately sort propositions will not be able to renounce it. The dogmatist's situation will progressively worsen over time, unless he or she makes lucky commitments. Ranging over possible dogmatists, what's more, we can see that being lucky has a low *objective* probability. This is clearly the case if we assume what most of us seem to, namely, that there are more possible unreliable methods, procedures, and so forth, than there are reliable ones. We could classify theories as true or false on the basis of coin-flipping, the reading of tea-leaves, the reading of palms, astrological charts of their advocates, etc. In searching for the reliable, we are looking for a needle in a haystack.

So the significance of justification in the externalist's sense of being (highly) reliable is simply that justified belief-forming processes will (much) more often issue in true beliefs than false beliefs. If given the choice between accepting a set of beliefs at random and accepting a set of beliefs derived from a highly reliable process, say, it is plausible that we should do the latter. We like reliability simply because we like truth. And as I will argue in Chapter 7, it is hard to see why we should expect science to rule out false theories rather than true ones if we do not possess a reliable source of observation statements.

It is worth adding that at one point, at least, Popper (1974a, p. 1114) appears to endorse the view that reliability of observation statements is important:

> Our experiences are not only motives for accepting or rejecting an observational statement, but they may even be described as *inconclusive reasons*. They are reasons because of the generally reliable character of our observations; they are inconclusive because of our fallibility.

The fallibility arises on two counts. First, we may be wrong that our observations are reliable. Second, our experiences can lead us to form mistaken beliefs. But the possibility of failure in the quest for reliable sources of information does not mean that the quest for such sources is futile, any more than the possibility of failure in the quest for truth means that the quest for truth is futile.

6.1 A Virtue Epistemological Approach?

The critical attitude may also be defended in a related but subtly different way, namely, by advocacy of the view that being critical is *virtuous* (in the epistemic, but not moral, dimension). Following Sosa (1980), one may suggest that justified beliefs are those beliefs which are grounded in intellectual virtues. One may then say that the primary intellectual virtue is being critical. In short, I take it that critical rationalists have no objection to, and indeed advocate, the view that being critical is being virtuous. So if we then say that being justified *just is* being virtuous, critical rationalists might concede that justification is important after all.[38] We will come on to discuss precisely what it takes to be a (good) pancritical rationalist in the final section of this chapter. For now, let us focus on the consequences of this idea.

Most epistemologists would probably want to agree that being critical (in something like the precise sense explained subsequently) counts as being virtuous. However, it is not so easy to see how possessing such a virtue contributes towards achieving what most take to be the appropriate aim in the epistemic dimension (i.e. the realm of inquiry and/or the realm of believing), namely truth. For example, David (2001, pp. 151–152) considers a variety of approaches to epistemology and concludes that in each: 'Truth is either explicitly referred to as a goal or aim, or it is implicitly treated as such'. And Alston (1989, pp. 83–84) is typical in suggesting that:

> Epistemic evaluation is undertaken from what we might call "the epistemic point of view" . . . by the aim at maximizing truth and minimizing falsity in a large body of beliefs.

If one insists on avoiding talk of belief, as Popper (1972) did in his proposal that we consider knowledge to be objective, then one may instead discuss 'a large body of hypotheses', 'a large body of theories', or even perhaps 'a large body of sentences'. But this is a sideshow. The pressing question is whether having a critical attitude and/or adopting critical procedures is either necessary or sufficient for achieving the aim of maximizing truth and minimizing falsity.

I do not think that it is sufficient, as we can see by considering a critical procedure that relies on a source that provides false basic statements more often than not. In that way, we might classify lots of true universal statements as false, and *vice versa*, and therefore worsen our epistemic lot. Admittedly a pancritical procedure would have wider scope than given in this example, but all criticism presupposes some 'basis' (even if it is temporary). The point is that shifting around via different bases in no way suggests that one will latch on to a *suitable basis* for criticism with respect to the aim.

This said, however, having a critical attitude (and aligned critical procedures) may, at least, be good enough for isolating inconsistencies and thereby ruling out some false beliefs (or conjunctions of sentences) provided that we take deductive logic (and that we can apply it reasonably well) as given. So it may, at the very least, fulfil the rather less lofty aim of *avoiding some falsities*. If we want to achieve the further aim of minimizing falsity and maximizing truth, then we will need rather more than (pan)critical rationalists are liable to think we can show that we have available! (One falsity may simply be replaced by another, or even several others, when ruled out. So avoiding some falsities need not minimize falsity.) We will return to this issue when we discuss the aim of science via an analogy with evolution, in Chapter 7. At the risk of spoiling the surprise, I will there argue that (pan)critical rationalists should reject the notion that the aim of science is truth, and can only defend the view that the aim of science is to rule out false theories *if they accept that we have a reliable means of isolating true observation statements.*

7. ON BEING A PANCRITICAL RATIONALIST

On the face of it, what's involved in being a pancritical rationalist is clear. A pancritical rationalist is willing to subject any of her beliefs to criticism, willing to give any of them up in principle, and so forth. On closer scrutiny, however, this description is undesirably vague. Clearly one need not be willing to criticise one's belief that being shot is often fatal while in the middle of a fire-fight. And when precisely it would be appropriate to renounce the belief that 2 + 2 = 4—how 'in principle' it would be reasonable to do so—is far from obvious. In closing this chapter I will explore these problems and endeavour to provide a clearer specification of what being a pancritical rationalist involves.

7.1 Bartley's Description of the Pancritical Rationalist

While not making any explicit attempt to *define* what it is to be a pancritical rationalist, Bartley nevertheless lists many properties that he thinks a pancritical rationalist should possess. A pancritical rationalist is:

> one who is willing to entertain any position and holds *all* his positions, including his most fundamental standards, goals, and decisions, and his basic philosophical position itself, open to criticism; one who never cuts off an argument by resorting to faith or irrational commitment to justify some belief that has been under severe critical fire; one who is committed, attached, addicted, to no position. (Bartley 1984, p. 118)

Unfortunately, however, Bartley here asks too much in some respects, and too little in other respects. In order to see this, let's begin by disentangling the requirements for our pancritical rationalist:

(1) She is willing to entertain any position.
(2) She holds all her positions—where 'position' includes standards, goals, decisions, and philosophical theses—open to criticism.
(3) She never cuts off an argument by resorting to faith or irrational commitment to justify some belief that has been under severe critical fire.
(4) She is committed, attached, addicted, to no position (where 'position' is understood in the same way as specified in proposition (2)).

I will begin by examining each requirement in turn, in a separate subsection. After moderating each as necessary, I will then consider if the resulting requirements need to be reinforced.

7.2 'She is willing to entertain any position.'

My phone rings as I am busy writing this section—I normally work from home—and I answer it with mild irritation. I don't recognise the voice of the person on the line, and they ask me to confirm my name. I confidently believe it's a marketing call, consider this to be an invasive activity, and therefore hang up. I'm simply unwilling to entertain the possibility that it is not, *in the sense of treating it as a live option*, because this will be time-consuming and I judge the probability to be low.[39] (Clearly it is often reasonable to act on a belief or other position while accepting that it might be wrong; that is not the issue here. The problem, rather, is whether it is ever acceptable to avoid entertaining the possibility that it *is* wrong, and if so when.) Have I failed to be a good pancritical rationalist?

Perhaps I have failed. It could have been a medical emergency (although I assume the caller will try again in that event). But imagine now that I instead confirm my name, wait, and am told by the person that they are indeed trying to sell me something, namely double glazing. In terms of 'positions', in the broad sense discussed earlier, we may say that the marketer is inviting me to consider my intention not to buy double glazing, my decision to spend my income on things other than double glazing, my goal to finish writing this section today rather than consider domestic issues, and my philosophical belief that it is a minor ethical transgression to call people in their homes in order to try to sell them things (in this sort of instance, e.g. when the call isn't being made under duress). If that's what it takes to be a pancritical rationalist, however, then I am afraid that I cannot even come close.[40] Moreover, I do not *want* to come close![41] I want to tell the caller that they should consider getting another job if they can—one that doesn't involve bothering other people with unsolicited and unwanted calls—and to delete my telephone number from their company's records. I then want to think no more about it.

Now the positions that I am apparently unwilling to consider, in this example, are remarkably mundane. They are nothing so grandiose as fundamental philosophical theses, or fundamental standards or goals. (Admittedly, they might be based on fundamental standards or goals, e.g. my work ethic, and there may be some basic ethical principles, utilitarian perhaps, that underpin my belief about the undesirability of cold-calling. But they might stand or fall irrespectively, on reflection.) Requirement (1) is therefore completely unrealistic, at best, and is plausibly undesirable even as an ideal to strive towards. I could get closer to that ideal by simply considering just one relevant position, e.g. whether I might benefit from having my house double glazed in the near future, in response to the cold call. But I don't even want to do that. Not now! Moreover, I am not even willing to set aside time to entertain the position in the future.

The basic strategy for solving this sort of problem with requirement (1) is clear. What's needed is something like:

(1*) She is willing to entertain any position *under appropriate circumstances.*

The problem with this replacement, however, is that it is undesirably vague. The worry is how we specify precisely when it is appropriate to refuse to entertain a position and when it is not. What exactly counts as an *appropri-*

ate circumstance? We can start by saying that we are interested in circumstances of an epistemic flavour; so the entertainment in question should be as a result of epistemic stimuli, rather than anything else. Being willing to think carefully about whether I want to convert to Islam while a fundamentalist's knife is at my throat may be wise, yet is not relevant to the problem at hand.

But going back to the cold-calling example, by way of a thought experiment, let us imagine that God suddenly declared to me that I would have all the time I liked to think over my positions relating to this call, with no ultimate loss in expected utility whatsoever (e.g. no loss from point of view of opportunity cost), if I so desired. It is now considerably more plausible that I would acquire an epistemic responsibility (at least) to take Him up on His generous offer. Just as I am obliged to take up free true information, according to the result of Ramsey (1990) and Good (1967) that one can never diminish expected utility by so doing and Carnap's (1962) principle of total evidence, so I am obliged to take up free reflection time (provided that there's no disutility in reflecting) according to pancritical rationalism.[42] In fact, the value of free reflection would be derivative from the value of gathering true information (and/or replacing false information), if reflection were to have a probability of unity of resulting in the acquisition of additional true beliefs (and/or a loss of false beliefs). However, high probability (coupled with a low probability of making adverse belief changes) will suffice from an externalist perspective. Taking up free reflection would then be a good strategy in the long run. (In this argument, I assume epistemic consequentialism, i.e. roughly that the epistemic status of a cognitive act is determined by the value of its consequences; but other arguments, e.g. concerning what counts as virtuous epistemic behaviour or doing one's epistemic duty, may be employed by those who disagree with this thesis.[43])

It may be added that reflection would also be valuable merely if it resulted in a non-zero probability (or even possibility, if zero probability does not preclude possibility) of isolating inconsistencies in one's belief set. A free chance to do this, and to restore consistency (or move closer to it), should not be forsaken.

This is an improvement, perhaps, but it would be nice if we could say rather more; after all, it is far from obvious that one should choose to reflect on just *anything* simply because one has the opportunity to do so for free. Given the opportunity to reflect freely on whether 1 + 1 = 2, for example, should I really take it up irrespective of the context in which I find myself? And what if I am given the choice of either reflecting on whether 1 + 1 = 2 or on foundational issues in contemporary physics? Isn't it fair to say I should choose the latter?

Inspired by Popper's emphasis on problems and problem-situations,[44] one might therefore be tempted to suggest that the pancritical rationalist should be willing to consider a position only when it is (or seems) relevant to some problem at hand, or is (or seems to be) an alternative to one of one's positions which has itself (apparently) become problematic.[45] (In the previous example, we may say that contemporary physics is more problematic than 1 + 1 = 2.) From an internal perspective, there appears to be something wrong in refusing *come what may* to consider a belief that one recognises is rendered problematic by one of one's other beliefs. (A simple case would be where a probabilistic theory is challenged by frequency data. Refusing to consider the possibility that a die is biased after seeing forty 'five' results in one hundred rolls, when lacking any other data on rolls of the die, may be an appropriate example.) From an external perspective, there is also something wrong with someone who has inconsistent beliefs.

From either perspective, there are important limitations on (1). First, one may be limited in principle, and not only in practice, as to what one can perceive to be problematic. Second, one may be limited in principle as to what kind of 'basic beliefs' one can attain. Thus some beliefs (e.g. religious ones) may be impossible to render problematic from an external perspective (at least if inconsistency with observation is all that is demanded). Either way, the notion of 'problematic' becomes central and requires careful refinement. This is not the present project, however. Suffice it to say that a position's relevance to some problem-situation is necessary, *but not sufficient*, for 'appropriate circumstances' to occur.

7.3 'She holds all her positions—where "position" includes standards, goals, decisions, and philosophical theses—open to criticism.'

To some extent, this requirement meets objections similar to those raised against (1). Think back to the example of the cold call. Just as one might say that I would be wrong to refuse to entertain the position that I should buy the double glazing, so one may say that I would be wrong not to hold my position that I should not buy the double glazing open to criticism. (Furthermore, one might add that I should hold my position that *I should not bother to entertain buying the double glazing* open to criticism.) Requirement (2) should therefore be replaced with:

(2*) She will hold any of her positions—where 'position' includes standards, goals, decisions, and philosophical theses—open to criticism *under appropriate circumstances*.

There is not a complete overlap between (1*) and (2*), however, because sometimes one might refuse to entertain a position that would not require revision (or abandonment) of any of one's current positions if it were to be accepted. One might refuse to consider theories concerning black holes simply because one has no interest in astronomy (and has no theories about black holes at all). Thus propositions (1*) and (2*) are interestingly distinct.

But can a wedge also be driven between the notion that one should hold a position open to criticism and the notion that one should entertain alternative positions? A simple way to discuss this is to consider negations, e.g. 'God exists' versus 'God does not exist'. Can a strong believer in the former hold it open to criticism without entertaining the latter? I believe that the answer lies in the affirmative, at least for some reasonable interpretations of 'criticism' and 'entertainment'. A strong believer may, for instance, be highly confident that no criticism of 'God exists' would ever succeed, and fail altogether to 'entertain' the possibility of His non-existence except in so far as he will devise arguments intended to show that said possibility does not obtain. So while said believer would be happy to employ an argument with 'God does not exist' as a premise, he would do so only with the aim of arriving at a *reductio ad absurdum.* In short, his position would be that 'God exists' is open to criticism, but never successful criticism. He would hold it open to criticism while failing to entertain its negation.[46]

Is the opposite true? Is it possible to (seriously) entertain a position contrary to another that one already holds while failing to hold the latter open to criticism? Again, I take the answer to lie in the affirmative. Someone may believe that 'God exists' and 'God does not exist' are not the sort of claims that that can be (genuinely) criticised—e.g. due to a commitment to mysticism—while nevertheless accepting that they are meaningful. This person may seriously entertain the idea that 'God exists' is false from time to time, in response to changes in personal emotional state, but nevertheless maintain belief in God's existence.[47]

I should also draw attention to an additional change of wording between (2) and (2*). To require the pancritical rationalist to hold *all* her positions open to criticism may be thought to also require that she do so for all positions simultaneously, which is naturally to require the impossible. It is enough, rather, for her to hold any of her positions open to criticism; either in isolation or as logically related groups. She cannot hold her position that 'England has a capital city' open to criticism without also holding her position that 'London is the capital city of England' open to criticism, at least if we assume that she has a reasonable grasp of logic, after all.

Finally, note that (1*) and (2*) can come together in some cases. Tim Williamson mentioned to me an acquaintance of his that would refuse to discuss certain of his religious positions because he feared that he would come to doubt them and change his mind. (He may have some 'meta-position' that changing his mind would result in a great personal penalty, of course. But presumably that 'meta-position' would also change if certain positions, e.g. 'God exists', were renounced.) This person accepts that his religious positions are criticisable, and even entertains the possibility that they are successfully criticisable. But he refuses to 'hold [many of] them open to criticism'. It is plausible, moreover, that he also avoids entertaining alternatives for fear of changing his mind. In this case, in short, the failure to satisfy (1*) seems related to the failure to satisfy (*2). (Perhaps this is due to the meta-position, and the resultant fear.)

This case also raises a further interesting issue, however. Might we not say that this person is *deliberately preventing his positions from becoming problematic*, and thereby doing his best to ensure that (what we have previously taken to be) a necessary condition for 'appropriate circumstances' never occurs? In fact, (2) *does* appear to forbid such behaviour. But a consequence of introducing the 'under appropriate circumstances' clause in (2*) is that such behaviour is no longer forbidden. This is undesirable; we will return to the issue later.

7.4 'She never cuts off an argument by resorting to faith or irrational commitment to justify some belief that has been under severe critical fire.'

Clause (3) is less objectionable than either (1) or (2), so requires considerably less discussion. The intent behind it is clear: it is inappropriate to terminate a critical discussion of some position—although Bartley uses only 'belief' in this clause, perhaps as the result of an oversight—simply by reaffirming the position and ignoring the criticism presented. I experience this sort of behaviour not infrequently, especially in discussions with my mother. When she tires of listening to my criticisms of one of her positions, she often says "Well, that's what I think!" and refuses to engage with any of the criticism. There does, indeed, appear to be something wrong with that (above and beyond the fact that it infuriates me)!

The use of 'justify' in (3) may, however, raise an eyebrow. In the way that mainstream epistemologists use the term, one may say that to appeal to faith or irrational commitment in defence of one's belief (or position) is *not* to justify said belief (or position). This suggests a further moderation of (3). So we should prefer:

(3*) She never cuts off an argument by resorting to faith or irrational commitment in an attempt to justify (or even merely defend) a position that has been under severe critical fire.

It should also be emphasised that cutting off an argument is not, itself, unreasonable. One is not under any obligation to continue to discuss some position for as long as any critic of that position happens to desire, for instance. The point remains, however, that it is possible to terminate the discussion not by appeal to irrational commitment but instead by simply saying that one does not wish to discuss the position any further. Accordingly, one should also not 'tell oneself' that it is OK to ignore the criticism that has been presented *just because* it threatens some cherished position. There is an analogy, here, with Popper's prohibition on introducing *ad hoc* hypotheses (when these are conceived of as adjustments designed *purely* to save some theory), which we will discuss in further depth in Chapter 5. The way that astrologists dealt with the discovery of Pluto is a case in point; they merely accommodated it by saying that the position of such a planet only makes a difference over long timescales. And now that Pluto has been declassified as a planet, presumably they will say its position is irrelevant (or accommodate the existence of other dwarf planets such as Eris and Ceres)!

7.5 'She is committed, attached, addicted, to no position.'

Requirement (4) is a reasonably uncontroversial requirement (from an epistemic perspective) when it is explained that being convinced that something is true *is* acceptable, but being committed *to* its being true is not.[48] Being committed to articulating, defending, or exploring some position or theory is, of course, also acceptable.[49] The point is simply that one should be open-minded about the possibility that said position or theory is false, or wrong, or otherwise inappropriate.

We might ask, however, whether acceptance of (4) makes (3*) unnecessary. If one were attached to no position whatsoever, then why would one cut off an argument by resorting to faith or irrational commitment in a way that was epistemically blameworthy? One could certainly cut off an argument *by appealing* to faith or an irrational commitment that one did not in fact possess, but what, precisely, would be the harm in that (provided one had a good reason to want to stop the argument, e.g. severe boredom with discussion over trivial detail)? It would not necessarily do any harm to oneself, epistemic or otherwise, relative to the available alternatives. Nor would it necessarily do any harm to the other participant in the discussion, especially if one did not have any responses to the criticisms advanced against the position (or even any further criticisms of the position). Indeed, one might even terminate the discussion on some position by appeal to faith and walk off with the new opinion that the position is false. Failure to share the opinion reached as a result of the discussion may be wrong on ethical grounds, but it is not an action forbidden by any of the alleged canons of *individual* rationality under discussion here.

My conclusion is that (3*) is therefore unnecessary, and that it is preferable to remove this rather than (4) because the latter requires no alteration. The discussion of (3*) is still valuable, however, in so far as this may be an important requirement for a pancritical rationalist *working in a collective of inquirers*. For one thing, failure for a group with shared interests to agree on probability assignments may result in a Dutch Book being made against it.[50]

7.6 A Summary of Findings and a Strengthening Addition

We have now seen that the following are true of a critical rationalist:

(1*) She is willing to entertain any position *under appropriate circumstances.*
(2*) She will hold any of her positions—where 'position' includes standards, goals, decisions, and philosophical theses—open to criticism *under appropriate circumstances.*
(4) She is committed, attached, addicted, to no position—where 'position' is understood in the same way as specified in proposition (2*).

We have also noted that:

(5) For 'appropriate circumstances' to occur in (2*), with respect to any peculiar position, it is necessary (but not sufficient) that the position has been rendered problematic.
(6) For 'appropriate circumstances' to occur in (*1), with respect to any peculiar position *that precludes a currently held position*, it is necessary (but not sufficient) that the position has been rendered problematic.

Finally, as we saw in our discussion of (2*) with reference to Tim Williamson's acquaintance, we must close a resultant loophole, roughly, as follows:

(7) She does not intentionally prevent any of her positions becoming problematic.

Unfortunately, however, introducing (7) raises a further problem because it does not always appear to be unreasonable to prevent one of one's positions becoming problematic. Consider being under fire as an army officer, and having to make a quick decision about the appropriate tactics to adopt in order to win the fight. This is not a time to consider whether your actions leading up to this point have been satisfactory, whether you are worthy of command, or to reflect on the reasons for which your government has dictated that you must fight. To do your duty with a regard to the lives of the men under your command is of paramount import.

An appropriate reformulation of (7) may therefore be:

(7*) She does not intentionally prevent any of her positions becoming problematic *come what may.*

Admittedly, this is far from ideal in terms of precision. A useful artifice, however, may be to consider how the person *would* behave if they were given the possibility of allowing a position to *possibly* become problematic 'for free'. (The idea would be that if they didn't take up the offer, then they wouldn't be a pancritical rationalist. So again, we are dealing with a necessary but not sufficient condition.) The only worry with this strategy is that it appears hard to see why one should allow any of one's positions to become problematic when it may very well be appropriate, correct, or true (in the case of a belief) irrespective of becoming problematic.

Now if desired, we can imagine assuring the person that any position would only become problematic (when subjected to the free possibility of becoming problematic) if there were, indeed, something wrong with it (e.g. in the case of a belief, if it were false). Think back, now, to the person who does not wish to discuss his religious beliefs for fear that he will change his mind. If we assured him that his beliefs would only be rendered problematic if they were false, then he ought to be willing to discuss them. The worry now, however, is that the acquaintance of Tim Williamson already meets this condition but is still doing something wrong.

We may circumvent this worry by instead relying on the idea, discussed earlier, that any reflection (e.g. 'free reflection') will reliably lead to an improvement in epistemic state (such as a gain in true beliefs, a loss in false beliefs, or the removal of inconsistent beliefs), and *that the person under consideration has been assured that this is true*. Then letting any position become problematic would be desirable on epistemic grounds (from either an externalist or internalist perspective).

This concludes our discussion. It may emerge that there are errors in this account of what it is to be a pancritical rationalist, and it is neither as precise nor as complete as I should like, but it is nonetheless more precise and comprehensive than any currently available alternative.

ACKNOWLEDGEMENTS

Parts of this chapter are based on Rowbottom and Bueno (2009) and Rowbottom (2005).

2 Induction and Corroboration

Popper is perhaps best known for his anti-inductivism, if not for his emphasis on falsification; for his view that the success of science does not depend on inductive inferences (because the scientific method does not require them).[1] For Popper, what matters in science is what we do with theories when we get them (or the so-called 'context of justification') rather than how we come up with those theories in the first place (or the so-called 'context of discovery'). In short, it is irrelevant whether one arrives at a theory by drinking several pints of beer and being very imaginative or by thinking in an inductive way (consciously or unconsciously), e.g. by generalizing from a small number of instances. Origin is irrelevant to judging the worth of a theory. What matters are features such as (empirical) accuracy, internal consistency, and scope (or what Popper calls 'empirical content'). And whether a theory possesses these can be judged when it is already on the table.[2] In Popper's own words:

> The question how it happens that a new idea occurs to a man—whether it is a musical theme, a dramatic conflict, or a scientific theory—may be of great interest to empirical psychology; but it is irrelevant to the logical analysis of scientific knowledge. (1959, pp. 30–31)

Naturally, one might criticise one's own ideas before publicly expressing them (whether scientist or not); and indeed, it is often wise to do so. But this process should not be confused with the idea's *genesis*, or the process by which it is brought before the (conscious) mind for the first time. It is this genesis that Popper suggests is epistemologically irrelevant. Using an apt academic analogy, Miller (1994, ch. 1) avers that we should have no entry requirements, but should have stringent and rigorous expulsion procedures.

In fact, Popper thought that *no form of ampliative (i.e. non-demonstrative) inference whatsoever* is required for good science. Neither simple enumerative induction nor sophisticated inference to the best explanation (as championed, for instance, by Psillos 1999 and Lipton 2004) is needed in order to make progress (or even in order to improve efficiency). Deduction suffices.

However, it is not entirely clear how this position relates to critical rationalism, i.e. the view that we should adopt critical attitudes and corresponding critical methods in inquiry. Strictly speaking, one might adopt a highly critical attitude yet nevertheless make (and believe that one ought to make) inductive (or other ampliative) leaps. And furthermore, *such leaps are permissible—even if unnecessary—according to Popper's view as articulated earlier.*[3] Indeed it might even be suggested that ampliative inferences can perform comparative critical functions, e.g. that one could put all 'live' theories on the table and select 'the best' on the basis of which one had the most explanatory power (i.e. by inference to the best explanation).

In what follows, we will try to unravel the relationship between critical rationalism and anti-inductivism (or, to coin an expression, anti-ampliativism). We will see that what we take the aim of inquiry to be is crucial; for although some ampliative moves may favour theories which are preferable on pragmatic grounds, it is not convincing that they favour true (or highly truthlike, or the more truthlike) theories.

1. CRITICAL RATIONALISM AND INDUCTIVISM

Let us start by considering (pure) inductivism, which is broadly the view that we should start with an open mind, gather data, and then prefer those theories which are best inductively supported (e.g. most raised in probability) by that data. Some have associated this view with Bacon, although perhaps somewhat unfairly.[4] There are naturally many arguments against pure inductivism, most notably that starting 'with an open mind' cannot mean starting without theories, because then one would be incapable of gathering any data! As Popper (2002 [1963], p. 61) put it:

> [T]he belief that we can start with pure observations alone, without anything in the nature of a theory, is absurd . . . Twenty-five years ago I tried to bring home the same point to a group of physics students in Vienna by beginning a lecture with the following instructions: 'Take pencil and paper; carefully observe, and write down what you have observed!' They asked, of course, what I wanted them to observe. Clearly the instruction, 'Observe!' is absurd . . . Observation is always selective. It needs a chosen object, a definite task, an interest, a point of view, a problem. And its description presupposes a descriptive language, with property words; it presupposes similarity and classification, which in their turn presuppose interests, points of view, and problems.[5]

But put this objection, and others of a similar ilk, to one side. Imagine that there were no barriers to proceeding in this way: that we could patiently make many theory-neutral observations, and only eventually use these to select a theory. What would be wrong with doing so?

The short answer is that it would be entirely uncritical. No-one—not even the most ardent champion of induction—seriously believes that adopting such a procedure would guarantee the truth (or truth-likeness) of the resultant theory. The truth of a claim such as 'All rabbits are brown' is not entailed by the observation of a trillion brown rabbits (and observation of brown rabbits without exception) any more than it is by the observation of a single brown rabbit. (More carefully, one should speak here of observation statements because observations never *entail* anything whatsoever.) This is because 'all' has maximal scope in scientific theories (or at least in certain crucial cases, e.g. classical mechanics and special relativity); it concerns all past, present, and future instances. So in short, 'all bunnies' means 'all bunnies in the actual world', 'all electrons' means 'all electrons in the actual world', etc.; and that's at the bare minimum. (On some accounts of laws, such claims must be true in all possible worlds![6])

So at best, such a procedure would only lead to theories which were probably true (or truthlike). (We will discuss how we should understand such a probability claim later; I will there argue that it is not interesting unless 'probably' is understood in an aleatory or chance-based sense, which it cannot reasonably be.) Therefore it would be wrong to devote no effort to testing any of those theories once they had been induced; to simply induce them and leave them be. (Naturally, testing those theories won't lead to certainty either. But that is not the point. What it will allow us to do is to correct mistaken inductions.) Criticising theories—by way of subjecting them to tests, both logical and empirical—is an important part of good science even if inducing theories is another significant part. There is a sense in which this is true even for Kuhn, as we will later see in Chapter 6, despite his emphasis on the importance of dogmatism, and the struggle to make facts fit theories, in science. Indeed, we will later see that a testing step is a feature of all the artificial intelligences that have, some allege, been successful in deriving scientific laws by mechanical induction.

Hence, one might say that we can and should adopt so-called 'inductive-deductive' methods; indeed, such an approach is recommended in many popular textbooks on research methods in the social sciences.[7] The basic idea is that one should use induction to come up with theories, and *only then* indulge in criticism. Such a two-stage process is good, one might suggest, because the theories put onto the table in the first place will less often be wrong than they might otherwise be (if they were selected, say, by pure chance). In the following section, however, we will see that this is plausibly wrong.

It seems fitting to close this section by adding that even deductive logic is only useful, for the critical rationalist, *as an organon of criticism*:

> We no longer look upon a deductive system as one that establishes the truth of its theorems by deducing them from 'axioms' whose truth is quite certain (or self-evident, or beyond doubt); rather, we consider a

> deductive system as one that allows us to argue its various assumptions rationally and critically, by systematically working out their consequences. Deduction is not used merely for the purposes of proving conclusions; rather, it is used as an instrument of rational criticism . . . (Popper 1983, p. 221)

2. CRITICAL RATIONALISM AND INDUCTION

Committing to belief in the conclusion of an ampliative inference (or preferring a theory selected by an ampliative procedure) is always risky, as we have already seen, even if one's premises are true beyond all possible doubt. But the advocate of induction as part of the scientific method may insist that it is less risky than relying on pure guesswork.

I would follow Popper, however, in denying that this is true. In fact, as I will argue in the following, it seems to me that to make an inductive inference is just to make a particular kind of guess (which is no better or worse than other kinds of guess). Like Popper, I have no problem with guessing in that kind of way. What I object to, rather, is the notion that one *ought to* guess in that kind of way.

It is interesting that van Fraassen (2007, p. 343), the architect of the stance view discussed in the previous chapter, agrees: 'I do not think that there is such a thing as Induction, in any form . . . *there is no purely epistemic warrant for going beyond our evidence*'. Settle (1990, pp. 404–405) summarises the critical rationalist position nicely:

> [Critical rationalists] think the point is that inductive inferences are not compelling, so that no one should be thought irrational who refrains from accepting conclusions from them. Other people think it irrational not to believe what induction supports, even though induction is non-demonstrative. I find it hard to sympathize with this latter view, hard to locate what it is about inductive arguments that warrants such a demand upon my allegiance. And I agree with Popper that the view seems mischievous. Why should I feel rationally compelled, as opposed to psychologically or physically constrained, to believe what may turn out to be false? And why rationally compelled, as opposed to invited or attracted? If the conclusions of non-demonstrative arguments with true premises could be false, would not that be a reason for refraining from belief, for exercising caution, especially if the price of wrong belief were high?

Settle is not entirely fair to inductivists, who might instead claim that accepting/believing the result of an inductive inference is simply less risky than the alternatives, or that only a peculiar degree of belief in the result of such an inference (given the premises) is required. (So if the probability of *p* given q is 0.99, then the appropriate degree of belief in p is 0.99 when q is assumed.) Nevertheless, he captures the essence of the critical rationalist's worry.

How might an advocate of ampliative inference in science respond? One general defensive line, inspired by Carnap (1968, pp. 265–267), is that deductive inferences cannot be satisfactorily defended without appeal to deduction, e.g. to a sceptic about deduction, any more than inductive inferences can be satisfactorily defended without appeal to induction.[8] (The same may be said of other forms of ampliative inference, if these are held to be significantly distinct from induction.) However, this is not really the issue. Deduction relies on the recognition that there are particular rules that govern propositions (or sentences). Whatever some proposition means, it cannot be true and false simultaneously. Nor can it be neither true nor false. And so on. More broadly, the notion is simply that if two descriptions are mutually exclusive then they cannot be applied simultaneously (and that any given description either applies or does not apply).[9] (This is true of prescriptions, too; classical logic is the foundation of standard modal logic.) As emphasised in the closing quotation of the previous section, deductive logic is therefore a tool to help us to spot, and eliminate, inconsistencies in our beliefs and theories (and/or resultant works such as monographs).[10]

Inductivists may also emphasise that deductive validity is purely syntactic, whereas its alleged inductive analogue is dependent on content (which is just as important). I do not, however, think that this is correct. In addition to strict logical necessity—namely, that which is necessary in virtue of the laws of logic alone—deductivists acknowledge narrow logical necessity. The latter is dependent not only on syntax, but also on the meaning of terms; the classic example is "All bachelors are unmarried men", where 'bachelor' means 'unmarried man'. And it seems obvious to any proficient English speaker that this statement is true.[11]

Compare this with "99 per cent of rabbits are brown. Tim is a rabbit. Therefore, Tim is brown". Are we to think that 'brown' is part of the meaning of 'rabbit'? I believe it is clear that it is not. 'Rabbit' and 'brown' are not coextensive, and it is perfectly possible to understand what a rabbit is without ever having experienced a brown thing. (Consider those who have been born blind. Surely we do not want to contend that they cannot grasp what a rabbit is, or that their grasp is partial?) In saying this, I recognise that some philosophers, such as Dretske (1977) and Armstrong (1983), might claim that brown (*qua* universal) is somehow related to the kind of rabbit. (Note that brown is the dominant colour for rabbits' fur.) But then we would be discussing *metaphysical possibility*, which is quite different. Deductivists will not accept that one can invoke metaphysics in order to defend induction without defending the metaphysics too. And an ampliative defence of a metaphysical position—e.g. appeal to best explanation to prefer a four category ontology such as that advocated by Lowe (2006)—will fail to satisfy! In the end, I think it is fair to say that most proficient English speakers comprehend, without any special philosophical training or thought, that "If you believe/assert that an entity is red all over, then you should not believe/assert that the entity is blue all over". This is because 'red all over' perspicuously precludes 'blue all over'. But "You should have a degree of belief of 0.01 in 'The next policeman you encounter will be corrupt' given only the information that '99 per cent of policemen are not corrupt'" is hardly as clear. It is highly controversial, *which suggests that it is not a matter of meaning.* This holds even if betting quotients are substituted for degrees of belief, as I've illustrated elsewhere (Rowbottom 2007b).

Nevertheless, arguments *for* the significance of induction are possible. First, one might say that some 'rules for guessing' are indispensable in so far as one must take account of science past in formulating new theories.[12] A simple example would be special relativity, according to which objects behave in approximately the same way as predicted by classical (Newtonian) mechanics provided that the relevant gamma factor is low, $\sqrt{1-\frac{v^2}{c^2}} \approx 1$, i.e. speeds sufficiently lower than the speed of light are involved. Isn't it important, or even crucial, for special relativity to have this feature?

A critical rationalist may agree that it is important for theories to explain how and why their predecessors were successful to the extent that they were (when appropriate predecessors exist). One of the tests Popper (1959, pp. 32–33) proposed for a theory on the table 'is the comparison with other theories, chiefly with the aim of determining whether the theory would constitute a scientific advance should it survive our various tests.' In his later writing, he fleshed this out as follows:

> [A] new theory, however revolutionary, must always be able to explain fully the success of its predecessor. In all those cases in which its predecessor was successful, it must yield results at least as good as those of its predecessor and, if possible, better results. Thus in these cases the predecessor theory must appear as a good approximation to the new theory; while there should be, preferably, other cases where the new theory yields different and better results than the old theory. (1981, p. 94)[13]

The crucial difference is between suggesting that one should worry about whether a theory fulfils such a condition *when generating it* and saying that it is sufficient merely to rule out a theory that does not fulfil the condition *after it has been proposed.* ('Rule it out' just means 'classify it as false'; it may still be used as the basis for developing similar, but strictly different, theories.) It should also be repeated that many scientists only propose theories publicly *after* they have performed the sort of test here discussed. That is to say, in developing a theory for public presentation a scientist may consider carefully whether it passes the test of explaining (many of) the successes of those theories it is intended to replace. If it does not, he may then abandon it, or tinker with it, and come up with something else to subject to the same sort of test. It therefore seems reasonable to conclude that heuristic-based advocacy of ampliative inference fails, at least if it is based on requirements like correspondence (i.e. explaining past successes).[14]

Nevertheless, second, preferring well-induced (or inductively highly probable) theories would be a good strategy if there were some link between choosing inductively highly probable theories and choosing true—or otherwise desirable, e.g. empirically adequate—theories. This view is associated with Reichenbach—see Gower (1997, ch. 10) and Popper (1959, §80) for criticisms[15]—and was shared by other leading (but less well-known) philosophers of science of the time such as Morris Cohen (1953 [1931], p. 130):

> It was a reasonable inference if it was the kind of inference that in an overwhelming number of cases leads to the truth if the premises are true. The probability of an inference, then, is the relative frequency with which its kind or type leads to true conclusions from true premises.

Nowadays, however, I suspect that several people attracted to probabilistic accounts of induction are attracted precisely because they equivocate on 'probability', and understand it both subjectively and objectively—or in both an epistemic and an aleatory fashion—in the same context.

The basic idea would be this: if you carry on favouring theories with high probability, then you'll almost certainly win in the long run. (This is essentially an *externalist* idea; that forming beliefs in an inductive way is reliable. As such, it meets the requirement of Bird [Forthcoming] that: 'any attempt to show that our inductive practices, whatever they are, can lead to knowledge will have to appeal to externalist epistemology in some form'.) The analogy would be with a game such as roulette, where the house always wins eventually (given a fair wheel) due to the odds offered on bets. A bettor will double her money if she successfully bets on black, for example. However, the probability of the ball landing on black is less than one half because less than half the numbers are black and the ball has the same chance of landing on each number. (Roulette wheels have one or two green numbers, namely 0 and 00.) So clients are offered odds that *would be fair if the chance of 'black' were the same as the chance of 'non-black'*.

The reason this idea doesn't work, in short, is that inductive probabilities don't appear to reflect chances. In order to see this, imagine the state of play in science at the turn of the twentieth century, and consider Newtonian mechanics—i.e. Newton's law of gravitation plus his laws of motion—which surely counted as 'well-confirmed' by almost any inductivist's lights, and therefore highly probable according to a probabilistic model of induction. For the sake of argument, let's say the inductive probability was 0.95 (which is presumably rather lower than some would say). Does this mean that in the majority of possible worlds with identical histories up to that point—95 per cent of those worlds, even—Newtonian mechanics is true (or truthlike, or even just empirically adequate)?

There are infinitely many such worlds—and this is a problem, as I will explain in the next chapter, for the logical interpretation of probability—so it is unclear how to construct a measure. There are multiple coherent options, and there is no obvious reason to prefer one over another. Even an appeal to the view that we should favour the most 'natural' measure—another idea we will discuss in the next chapter, with respect to Bertrand's paradox—does not help. No particular measure seems more natural than any other.

I suppose someone might object that we should consider metaphysical possibilities rather than logical possibilities. But first, this will be unsatisfactory to the extent that we must work out some relevant metaphysical possibilities on the basis of science. Second, even imagining the metaphysical possibilities are put beyond doubt, we may still be left with a world where Newtonian mechanics is true for every world where Newtonian mechanics is false. Or even worse, if strong determinism holds, it may be the case that the history of the world until 1900 is only the same (as it is in ours) in possible worlds with identical laws. So in that event the relevant probability of the truth of Newtonian mechanics is simply zero (provided we are right that Newton's laws don't hold in the actual world).[16]

One might also object that the whole idea of thinking in terms of possible worlds is incorrect in the context of considering objective probabilities, because these (e.g. construed as chances rather than frequencies) are world-bound. This doesn't, however, give any solace to those who would advocate an inductive strategy on probabilistic grounds. Even allowing for an inductive defence of induction—let's do this, for the sake of argument—what would need to be shown, empirically, is that well-confirmed theories do tend to have been true (or empirically adequate, or whatever) *and* that very highly confirmed theories have more often been true (or whatever) than theories that only ever achieved high confirmation values, and so on.[17] This hasn't been done, and there is every reason to suspect that it cannot be done.

Salmon (1990, p. 187), for example, suggests that:

> [P]rior probabilities [i.e. probabilities for hypotheses prior to data concerning those hypotheses being collected] . . . can be understood as our best estimates of the frequencies with which certain kinds of hypotheses succeed . . .

There are, however, a number of problems with this. First, we presumably want our prior probabilities to reflect actual frequencies (or even relative frequencies in the long run), but there is no reason offered to expect that 'our best estimates' will, in general, be anywhere near the actual values. Even accepting that we can give good estimates of frequencies of events in simple cases, e.g. when it comes to level roulette wheels and (unloaded) dice, it does not follow that we can do so in the case of complex hypotheses (concerning the motion of all objects, the space-time manifold, the behaviour of unobservable entities, and so forth). What, for instance, is the frequency with which a theory like general relativity succeeds? How does one estimate that?

Second, there is a problem concerning how we should classify hypotheses with respect to kind (i.e. how we should understand 'certain kinds'). Consider Newtonian mechanics (by which I mean, roughly, Newton's laws of motion plus Newton's law of gravitation). What kind of hypothesis is this? We could consider it as a member of the class of hypotheses *proposed by Newton*. Or we could consider it as a member of the class of hypotheses *concerning motion*. And then again, we could consider it as a member of the class of hypotheses *concerning physical objects*. Clearly we will have different priors depending on our taxonomy of theories, and there is not any privileged way to view matters. The only obvious way around the problem—which is fundamentally the old reference class problem that plagues the frequency interpretation of probability[18]—is to consider each hypothesis as entirely unique. But then we can hardly estimate accurately the probability with which such a specific hypothesis—one which we have never before encountered—succeeds! We could not have any data on that.

It is also dubious that the priors of many theories which were quite successful, such as special relativity, *were* high. So it cannot, therefore, be right that we should be guided about which theory to put on the table simply by a high prior; at best, it would have to be the highest prior (of some competing options) that would be our recommended guide.

Admittedly, some believers in the importance of induction in science might concede that we shouldn't select theories on the basis of their relative priors. And they might do this while maintaining that theories can nevertheless be confirmed—rather than merely corroborated, on the model we will examine in the following—by evidence and/or testing. *In short, this would be to concede the irrelevance of the 'context of discovery' while maintaining that induction is valuable in order to confirm theories after the fact of their proposal.* Such a person may be referred to the earlier discussion of how preferring highly probable theories only appears to be linked to achieving the aim of science (whether truth or something else) if the probability involved is objective.

It is also worth emphasising, as Keynes (1921, p. 76) noted, that there is no general link between weight of evidence (i.e. quantity of relevant information) and probability of success (in so far as truth is concerned) when one works within an epistemic theory of probability:

> Weight cannot . . . be explained in terms of probability. An argument of high weight is not "more likely to be right" than one of low weight; for the probabilities of these arguments only state relations between

> premiss and conclusion, and these relations are stated with equal accuracy in either case. Nor is an argument of high weight one in which the probable error is small; for a small probable error only means that the magnitudes in the neighbourhood of the most probable magnitude have a relatively high probability, and an increase of evidence does not necessarily involve an increase in these probabilities.

Keynes's point is that probability relations *are what they are*. (And this goes equally for conditional personal degrees of belief, under the subjective interpretation of probability). We must therefore be wary of adopting the seductive but false view that gathering more evidence somehow leads to 'the true probability' (and resultantly, perhaps, 'the truth') unless we are concerned with estimating aleatory probabilities (e.g. by flipping a coin).[19]

In fact, there are many cases where having all the available evidence would lead one to believe in a false hypothesis, whereas having partial evidence would lead one to believe in a true one. (This does not depend on 'available' being understood as 'available in practice', rather than 'available in principle'.) And why should we expect these cases to be less frequent than those where the opposite is true?

3. IMPROBABLE SCIENCE

Intuitively speaking—and I suspect that this is still the majority view—it is tempting to think of contemporary science as populated by *highly probable* theories. If we just think that probabilities reflect (coherent) degrees of personal belief, then no doubt this is true. You might, for instance, strongly (and coherently) believe in a consistent subset of the scientific theories that you were taught in school. And similarly, many expert scientists no doubt strongly (and coherently) believe in numerous theories in their respective areas of specialism. As it stands, however, this result isn't terribly interesting. So what if someone is, or lots of people are, pretty confident that some theory is true? Bruno de Finetti (1937), one of the architects of the subjective interpretation, himself emphasised that his aim was only to explain the *psychological* reasons for which agreement occurs.[20] This is hardly the stuff of the philosophy of science, one might opine! Instead, surely, we should be interested in how the evidence we possess bears on the theories we have (irrespective of how strongly anyone just so happens to believe in those theories). Plausibly, there's a fact of the matter about that no matter what anyone thinks, in the same way that the microphysical composition of the table in front of me is independent of any theories I, or anyone else, might have about it.

Popper (2002 [1963], p. 77) suggested that if we look at probabilities in this non-subjective way, we will see that the aim of science is *not* to attain highly probable theories, but rather to proffer '*explanations; that is to say, powerful and improbable theories*'. He later explained:

> This may sound paradoxical to some people. But if high probability were an aim of science, then scientists should say as little as possible, and preferably utter tautologies only. But their aim is to 'advance' science, that is to *add* to its content. Yet this means lowering its probability. (2002 [1963], p. 386)[21]

In fact, Popper offered a striking argument that the probability of universal laws is generally zero, at least when probability is understood in a *logical* way (as explained in the following).[22] It is crucial to contrast this logical way with the (merely) subjective way, mentioned earlier, where probability concerns degrees of belief which are 'rational' if and only if they satisfy the axioms of probability; see Ramsey (1926) and De Finetti (1937). The standard argument in favour of this subjective interpretation of probability proceeds via Dutch Book considerations—see Gillies (2000, pp. 55–65) for a presentation, and Hájek (2005) and Rowbottom (2007b) for challenges—but we will pass over this here because we will return to the issue of the interpretation of probability in the next chapter.

Keynes (1921, p. 11), the architect of the logical interpretation, suggested that probability '[i]n its most fundamental sense . . . refers to the logical relation between two sets of propositions . . . Derivative from this sense, we have the sense in which . . . the term *probable* is applied to the degrees of rational belief . . . ' So his position was that one's degree of belief in *a* given *b* (i.e. accepting *b* as 'background information') is rational only if it has the same value as the (abstract) logical relation involving *a* and *b*.[23] Entailment of *a* by *b* is just a special case of a probability relation between propositions, one where the relevant relation has the value of unity. (And similarly, if *b* entailed ~*a* then the probability relation would be zero.)

What Popper argued, *pace* Keynes, is that rational degrees of belief do not map onto those fundamental logical relations (which are nevertheless probabilities). Instead, he defined logical probabilities without any reference to degrees of belief at all (Popper 1983, p. 292):

> In . . . the *logical interpretation of probability*, '*a*' and '*b*' are interpreted as names of *statements* (or propositions) and
>
> $$p(a,b) = r$$
>
> as an assertion about the *contents* of *a* and *b* and their *degree* of *logical proximity*; or more precisely, about the degree to which the statement *a* contains information which is contained by *b*.

That's enough on the logical interpretation, for the time being. Let's now consider Popper's argument that the logical probability of a universal law is zero, which he summarised at one point as follows:

> [E]very universal hypothesis *h* goes so far beyond any empirical evidence *e* that its probability *p*(*h*,*e*) will always remain zero, because

> the universal hypothesis makes assertions about an infinite number of cases, while the number of observed cases can only be finite. (1983, p. 219)[24]

The best example offered by Popper (1959, pp. 372–373) involves fitting a curve to a finite number of points in a finite universe. The fundamental idea is that no matter how many points we have, there are still infinitely many curves which can fit them.[25] So although any new data point will rule out some curves—not conclusively[26]—it will not increase the probability of any remaining curve which fits all the data. It should be noted that this argument only works provided there is no background information which favours one curve rather than the other, since Popper seems to be relying—although not explicitly—on something like the principle of indifference, 'that *equal* probabilities must be assigned to each of several arguments, if there is an absence of positive ground for assigning *unequal* ones' (Keynes 1921, p. 42).[27]

I think I can strengthen Popper's case, however, by instead presenting an example which relies on an actual scientific theory (or theoretical framework), namely Newtonian mechanics. Imagine a universe much more limited than our own, which begins at $t_1 = -1$ and ends at $t_2 = 1$, and contains only a few bodies (including an experimenter). In order to test whether Newton's laws hold, the experimenter sets up the bodies so that the one she is going to observe should—according to her theories and the relevant auxiliary assumptions[28]—have no resultant force acting on it. (Let's imagine that as far as she's concerned, there is only gravitational force.)

Newtonian mechanics 'tells' her that the body should have a constant instantaneous velocity, but it's impossible for her to measure this directly (just as it is in our own world). So she decides to measure the average velocity repeatedly, over increasingly short periods, in order to give the law the best test that she can. If the average velocity varies, this will indicate that the instantaneous velocity has varied.

Let's imagine that she devises a way to perform one test between $t = 0$ and $t = 0.5$, another between $t = 0.5$ and $t = 0.75$, and so on, halving the period of time for each measurement. It therefore follows that she can perform an infinite number of experimental tests in principle. But how could any finite number, which is all that she could achieve in practice, actually increase the *logical* probability of Newton's laws?

Consider a family of counterhypotheses, similar to Newton's laws but with an addition: 'There are periodic fluctuations in velocity, which occur every *r* seconds and have a duration of *s* seconds, but which "see-saw" such that the average velocity over *s* is exactly as predicted by Newton's laws.' There are infinitely many laws in this family, and no finite data (of the form gathered by our experimenter) can lead to anything less than infinitely many remaining unfalsified. Now if we use the principle of indifference over these options at any point in time, we'll have to say that they are equipossible. (By requiring that *r* and *s* are rational in the family of counterhypotheses under consideration, the infinities are rendered denumerable.[29]) But this means that the probability of each is effectively zero.

One obvious criticism of the foregoing argument is that the principle of indifference is flawed; indeed, this is the objection of Howson (1973, p. 155) to Popper's original argument. (We will see why the principle is flawed in the next chapter.[30]) But even if we dispense with the principle, it is not plausible that any particular member of the family of counterhypotheses emerges as any more probable than any other. (Note that the member where *r* and *s* are equal to zero may be thought of as Newtonian mechanics.) Each counterhypothesis has the same form; only the values of the variables are different in each case. So even if one accepted that a criterion such as simplicity should be used to weight the counterhypotheses, it is implausible that any one is simpler than any other. (It is equally implausible, with reference to the theoretical virtues enumerated by Kuhn [1977, p. 321], that any counterhypothesis has more scope, consistency, or fruitfulness than any other.) In any event, why should we think that simplicity, say, is a guide to truth (or even just empirical adequacy)?[31]

Another objection might be that such forms of hypotheses are often ruled out, in so far as rendered highly improbable, by other scientific theories. However, one must then question the evidential grounds for those other scientific theories, and whether *they* are highly probable on the empirical evidence. Many of them will be also universal; and all will only have been subjected to a finite number of tests. One may therefore suggest that *their* probability is zero. To appeal to further scientific knowledge is just to shift the problem back one stage. At some point there will be a terminus (if we are not to go around in a circle): a universal theory which can only be evaluated against finite relevant evidence, and which may therefore itself be said to have probability zero.

Unlike Popper's curve-fitting example, mine does rely on an infinite universe. It is 'infinite with respect to the number of . . . spatio-temporal regions' (Popper 1959, p. 363), rather than the number of concrete entities (and/or 'distinguishable things'). But it concerns a cluster of actual scientific laws being tested in a universe *considerably* more limited, in scope and content, than our own. It therefore seems reasonable to conclude that there are actual scientific laws which have zero logical probability relative to any evidence that we can gather. That is, on the assumption (which I will later question) that there really are logical probabilities.

4. TESTING AND CORROBORATION: A BRIEF INTRODUCTION

It would be counterintuitive (and counterproductive) to suggest that we are irrational to believe in universal theories such as Newtonian mechanics, given science as is and has been, even if their logical probability is zero relative to any finite empirical evidence. Epistemologists may therefore be quick to suggest that Popper must have rejected evidentialism—roughly, the thesis that one should suspend belief about whether *p* or ~*p* if one has no evidence for or against *p*[32]—and we will touch on that a little later. What Popper did say for sure is that scientific theories can enter our corpus of *objective knowledge* despite their improbability.[33]

Popper therefore severed the link between how well a theory has stood up to tests (and so, as we will see, how 'believable' it is) and how probable it is. In particular, he argued that the logical probability of a theory is not identical to its degree of confirmation (or corroboration): '*degree of corroboration cannot be a probability*, because it cannot satisfy the laws of the probability calculus' (Popper 1959, p. 363).

Formally, Popper proposed the following function in order to measure the corroboration of a hypothesis *h*, given evidence *e* and background knowledge *b*:

$$C(h,e,b) = \frac{P(e,hb) - P(e,b)}{P(e,hb) - P(eh,b) + P(e,b)}$$

The numerator has intuitive significance: 'The support given by *e* to *h* becomes *significant* only when . . . $p(e,hb) - p(e,b) >> \frac{1}{2}$' (Popper 1983, p. 240). The denominator fulfils a normalising role; it serves to limit the values of $C(h,e,b)$ to a maximum of +1 and a minimum of −1, since $P(eh,b) = P(e,hb)P(h,b)$, and $P(e,b) \geq P(eh,b)$.[34] Thus, if *e* supports *h* (relative to *b*), C has a positive value, whereas if *e* undermines *h* (relative to *b*), C has a negative value. There is also a third option: if *e* is irrelevant to *h* (relative to *b*), then C is zero.

Now, according to Popper, the experiment which produced the Poisson bright spot (*e*) was so impressive because the result was 'unlikely' given the background knowledge beforehand (*b*), but was a consequence of Fresnel's wave theory of light (*h*) conjoined with such knowledge. The theory and background knowledge entailed a risky prediction, and one that proved to be successful.

The bright spot appears in the centre of the shadow cast when an opaque disc (or sphere) is illuminated, i.e. in a place that one would expect there to be a 'solid dark patch' on the basis of everyday experience of shadows. The story behind the spot is also rather entertaining. Here's a short version. Fresnel presented a paper on his wave theory of light for a competition. Poisson, who was on the judging panel, disliked the wave theory of light (instead preferring a corpuscular account thereof) and therefore sought to refute Fresnel's paper by showing that such a spot would have to appear where (he thought we knew) it should not. The story might have ended there if Arago, another judge on the panel, had not performed the experiment and discovered the spot. The story has a happy ending. Fresnel won the competition![35]

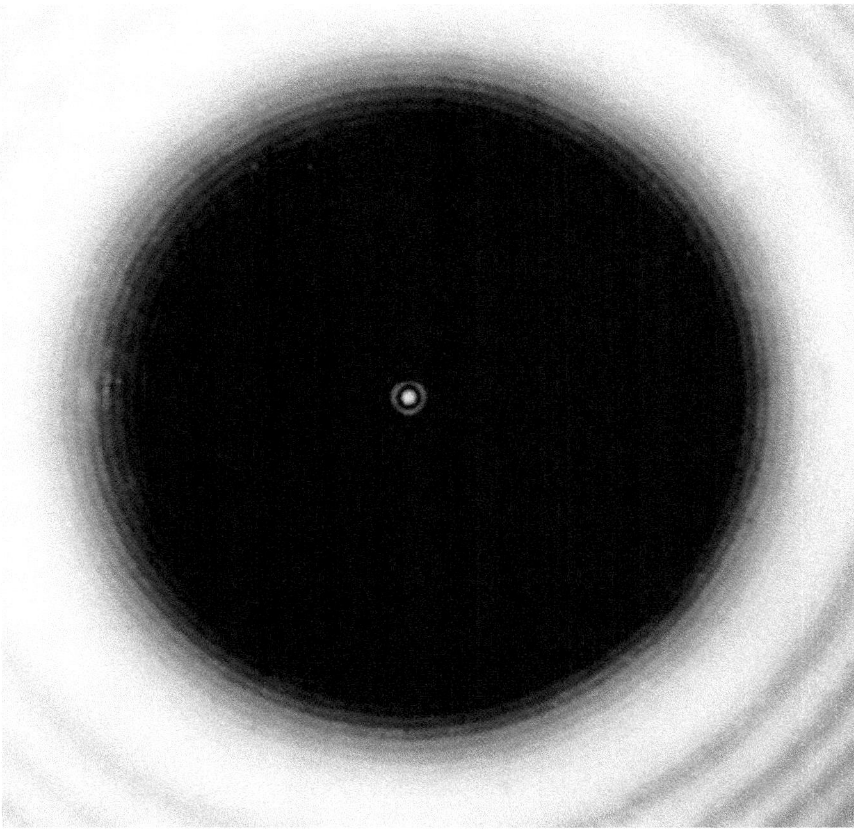

Figure 2.1 The Poisson bright spot.
Reproduced with permission from Rinard (1976, p. 70). Copyright 1976, American Association of Physics Teachers.

The history of science is replete with such episodes—e.g. Halley's prediction of the return of the comet which now bears his name, and the predicted time dilation of clocks on flying aircraft relative to clocks on Earth—in which a theory (or theoretical framework) survived a severe test, and in which the corresponding value of C was high. In general, C reaches its maximal value only under the following conditions: P(*e*,*hb*) = 1, P(*e*,*b*) = 0, and P(*h*,*b*) = 0. It reaches its minimal value only when P(*e*,*hb*) = 0; that is, when the evidence serves to falsify the hypothesis given the background knowledge. Popper (1983, p. 241) also suggested that 'the maximum value that C(*h*,*e*,*b*) can attain is equal to 1 – P(*h*,*b*) and therefore equal to the content of *h* relative to *b*, or to its degree of testability'.

We will cover corroboration in greater depth in the next three chapters and consider the proper interpretation of probability to employ in the formula, how we should distinguish between tests, whether it is correct to say

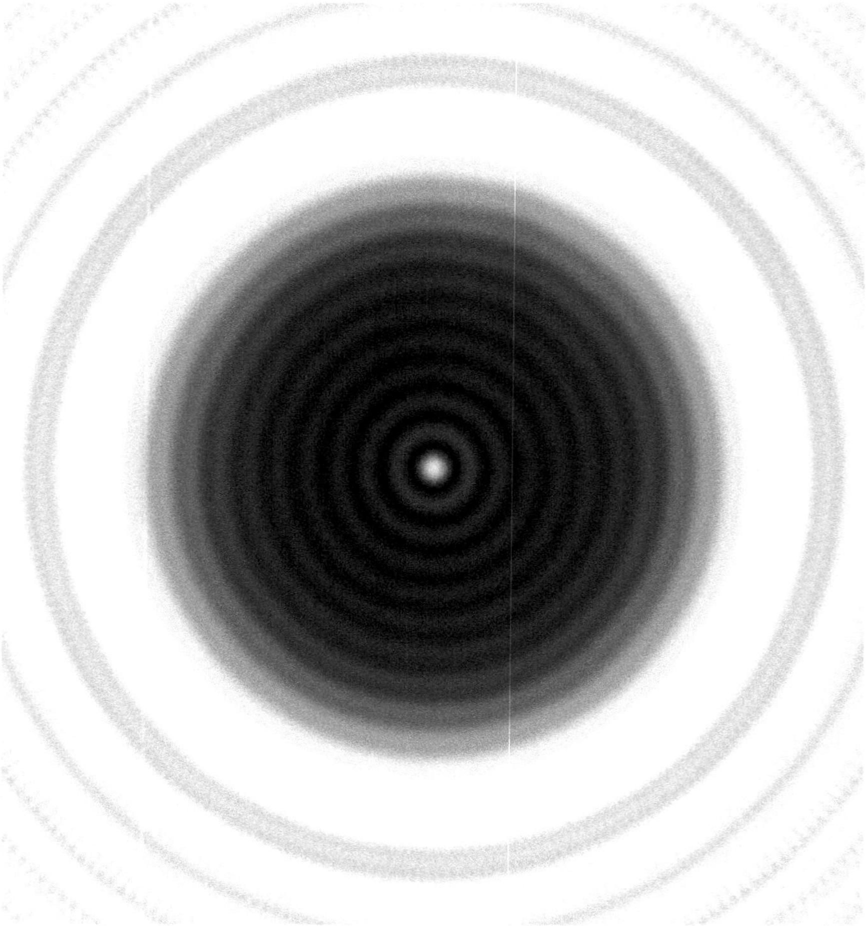

Figure 2.2 A computer-generated Poisson bright spot.
Produced using *Fresnel Diffraction Explorer* (by Dean Dauger)

that only prediction (and not accommodation of previous data) should be relevant, and how Duhem's thesis bears on falsification. In the following, however, we will focus on some objections to the renunciation of induction that may be tackled without considering corroboration in such depth.

5. OBJECTIONS TO THE RENUNCIATION OF INDUCTIVE METHODS

5.1. Why Believe in Spatio-Temporally Invariant Laws?

We have already seen that Popper believed that the logical probability of universal (spatio-temporally invariant) laws is zero. They say things not only about the present, but also the distant past and the far future. So one might wonder why we'd believe in them at all, if it weren't for induction! One might also wonder why we bother to look for them if there is no evidence that they exist.

The short answer to this, for the critical rationalist, is that it would be nice if they did exist. And if we were to find such laws, we would have a great deal of predictive (and retrodictive) power, not to mention a better understanding of our world. That, in essence, is why we look for them. (If they don't exist, after all, then what hope do we have for predicting the future? It would be defeatist to just assume that they don't.)

To do science we need not, however, *assume* that such laws do, in fact, exist. In the words of Miller (1994, p. 26), one may recommend the adoption of the 'methodological rule: to search for spatio-temporally invariant laws' without presupposing that the search will be successful. Or to put it more snappily: 'Scientific hypotheses propose order for the world; they do not presuppose it' (ibid., p. 27).

Rescher (1987, p. 126) has similarly suggested that metaphysical realism (i.e. the thesis that there is a mind-independent objective reality) is a 'postulation made on *functional* rather than *evidential* grounds'. And this is even more fundamental than the notion that there are spatio-temporally invariant laws (since such laws may also be understood, say, in a Kantian framework where transcendental idealism is coupled with empirical realism).

But to talk of posits (and searches) is not to talk of beliefs, and it is undeniable that most of us believe—particularly if we take our actions to reflect our beliefs—in spatio-temporally invariant laws. In the end, however, I think we should simply admit that many of our most fundamental beliefs are held on (utterly) non-evidential grounds. Unfortunately, in philosophy both past and present, the tendency is to refuse to accept such a conclusion and instead endeavour to defend the view that we know all those things that we intuitively think we do. (So it is assumed that we know there is a mind-independent reality, we know that we are not brains in vats, and so on, and so forth, and that these are just basic facts that require explanation.) I understand this. It is difficult to accept that many of our deepest convictions are without any evidential foundation; it is potentially damaging to our self-image as 'noble in reason [and] infinite in faculty' (Hamlet, II, 2). Yet acknowledge it, I believe, we should. Philosophy for the critical rationalist is about *testing* our intuitions and so-called 'common sense', after all!

It does not follow that our beliefs in universal laws are irrational just because they are not based on evidential grounds. On the contrary, our legitimate reasons for belief may be more diverse than evidentialists allow; they may be prudential or moral, for instance. I will not make it my business to offer a detailed argument against evidentialism because that would take us too far off-track—see instead Foley (1987, 1991), Owens (2000), and Booth (2007)—but will instead content myself with saying that if it is true, then the vast majority of our everyday beliefs (and our scientific beliefs in particular) are irrational. This seems absurd.

It is also worth emphasising that one may reject the very notion that one should believe *p* only when one has a reason to do so. Van Fraassen, for instance, defends the view that it is perfectly reasonable for us to believe anything that is not *rationally forbidden*:

> "[W]hat is rational is whatever is rationally permitted": *rationality is bridled irrationality.* (2004a, p. 129)

So if there is no evidence against *p*, then belief in *p* is permissible. In support of this epistemological voluntarism, as he calls it, van Fraassen appeals to the 'boringly repetitive failures of the idea of Induction and similar rule-governed concepts of rational opinion and its management' (2004b, p. 182). A critical rationalist would add little more than "Quite!" I leave the final word to Popper (2002 [1963], pp. 67–68):

> Hume was right in stressing that our theories cannot be validly inferred from what we can know to be true—neither from observations nor from anything else. He concluded from this that our belief in them was irrational. If 'belief' means here our inability to doubt our natural laws, and the constancy of natural regularities, then Hume is again right: this kind of dogmatic belief has, one might say, a physiological rather than rational basis. If, however, the term 'belief' is taken to cover our critical acceptance of scientific theories—a *tentative* acceptance combined with an eagerness to revise the theory if we succeed in designing a test which it cannot pass—then Hume was wrong.

The view of Popper (2002 [1963], p. 75), which we will explore in the following, was that:

> [O]ur belief in any particular natural law cannot have a safer basis than our unsuccessful critical attempts to refute it.

5.2. Rational Prediction

Salmon (1981) offers the sharpest formulation of the classic objection against a corroboration-based account of science. When we are faced with some pressing problem to solve, why should we prefer our best tested theories rather than others? Why, that is to say, should we care about corroboration values (except perhaps in so far as they sometimes indicate when a theory should be rejected)?

Assume it is clear that some theories should be avoided, perhaps because they have been classified as false due to their incompatibility with properly accepted observation statements (plus auxiliary hypotheses). Why should we not prefer the lesser tested of two competing, live, theories? And why should we not, time permitting, simply generate and act on an entirely new (untested) theory which is compatible with our observations (and auxiliary hypotheses) to date? In the words of Salmon (1981, p. 117):

> We ought not to employ premises which are known to be false if we hope to deduce true predictions. The exclusion of refuted generalisations does not, however, tell us what general premise should be employed. Typically there will be an infinite array of generalisations which are compatible with the available observational evidence, and which are therefore, as yet, unrefuted. If we were free to choose arbitrarily from among all the unrefuted alternatives, we could predict anything whatever.

Some critical rationalists appear to have suggested that there is no good answer to these questions, and that corroboration is unimportant in practical contexts (even if it remains important in theoretical contexts).[36] However, I think that much of our actual behaviour, theirs included, is inexplicable (or curiously biased) if this is true. Imagine, for instance, that you learn that you are suffering from a terrible, but curable, disease. Imagine, furthermore, that there are two theories about how to cure it but you can only act on one because you have little time left. One is tried and tested, and acting on this theory has always succeeded in curing sufferers from your affliction in the past. The other is brand-new and entirely untested. Which should you choose if your overriding desire is to live on (rather than to advance medical science)?

The point may be made that we should stick with testing apparently successful theories from the theoretical point of view of natural science—that we should not abandon a theory until it has failed—in order to ensure progress. But the scenario under discussion here does not concern a theoretical matter.[37] (Besides, it may also be helpful to test bold new theories even when there are well-established competitors.) It concerns which theories we should apply in order to further our practical ends. Why should we rely on the better tested theory *here and now*?

The answer has two parts. First, one may ask why we should expect acting on a theory to be successful in the future simply because it has been successful in the past. To this, the reply is just that there is no *evidential* reason to do so. Recall from the previous subsection that science proposes, but does not presuppose, that there are spatio-temporally invariant laws of nature. (We may indeed work with the hope that there are—and even the psychological constraint of deep-seated belief in—such laws, but this is irrelevant.) The point for present purposes is that we have to choose between proposals which are both universal.

Second, one may ask why one should expect the better tested theory to be more likely to be true, or truer, or even just (closer to) empirically adequate.

This is the key problem we will tackle in the next section, during the discussion of corroboration and verisimilitude. My rough answer, which I will develop there, is that the aleatory probability that the better tested theory would have been exposed as false (because empirically inadequate), if it were indeed false, is greater than the probability that the more poorly tested theory would have been exposed as false if it were false (given some plausible assumptions). In the case just discussed, the new untested theory has had no chance of being exposed as false if it is. The well-tested theory has had a significantly higher chance of being exposed, but has survived. *And even if this doesn't provide evidence that it is true, or highly truthlike, or even empirically adequate, it does provide a reason to prefer it.*

Note that the use of probability in this argument is in no way in conflict with the previous discussion. Recall that it was accepted, earlier in this chapter, that induction would be a good strategy if high inductive probability of a hypothesis corresponded to a high *aleatory* probability of its truth (or truth-likeness), e.g. if induced (or well-induced) theories were more often true (or otherwise epistemically praiseworthy) than guessed theories. It was argued that we have no grounds for thinking that this is true, although we will return to the issue when we discuss artificial intelligence in the final section of this chapter.

Before we continue, however, we should note that even theories which are widely accepted to be false can be, and regularly are, used for predictive purposes. Newtonian mechanics provides a fine case in point; we do not require relativity in order to consider problems in biomechanics, for example. Of course, Popper (2002 [1963], p. 74) recognised this:

> [F]alse theories often serve well enough: most formulae used in engineering or navigation are known to be false, although they may be excellent approximations and easy to handle; and they are used with confidence by people who know them to be false.

Miller (Forthcoming) also points out that it is often incorrect to say that we use well-corroborated theories to *derive* practical proposals. Rather, we tend to evaluate practical proposals by employing well-corroborated theories in a critical capacity. He writes:

> Our 'basis for action' . . . should be not the theory that has best stood up to criticism, since theories unaided do not make proposals for action, but the practical proposal that has best stood up to criticism, including criticism using the best tested theories available. The effectiveness of this best-criticized proposal may not be deductively connected to any theory in our possession.

Miller is undoubtedly correct about this, to my mind, but he sees this as a solution to the problem of rational prediction whereas I do not. He says, quoting from Popper (1974c, pp. 1025–1026), that we should adopt the practical proposal that best survives:

> the most testing criticism we can muster. *But such criticism will freely make use of the best tested scientific theories in our possession* . . . Why . . . does rational criticism make use of the best tested although highly unreliable theories? The answer . . . is exactly the same as before. Deciding to criticize a practical proposal from the standpoint of modern medicine (rather than, say, in phrenological terms) is itself a kind of 'practical decision' . . . Thus the rational decision is always: adopt critical methods which have themselves withstood severe criticism.[38]

However, the pressing question then becomes "Why should we consider it irrational to fail to use the best tested (or best corroborated) scientific theories in our possession for criticising practical proposals?" Why should we not use an untested theory in order to criticise a practical proposal? Return to the earlier example of the terminal illness. Now imagine that you have to choose between two different treatments. One way to determine which is best would be to look to a highly corroborated theory (for critical purposes). Another would be to use the sum of the numbers from the next national lottery to decide (e.g. if the sum of those numbers proves to be odd, then select treatment one; else, select treatment two). Why do the former, rather than the latter? To my mind, Miller has not, alas, satisfactorily answered this question.

In order to make this crystal clear, consider the following passage from Salmon (1981, p. 121):

> The question is not whether other methods—*e.g.*, astrology or numerology—provide more rational approaches to prediction than does the scientific method. The question is whether the scientific approach provides a more rational basis for prediction, for purposes of practical action, than do these other methods. The position of the Humean skeptic would be, I should think, that none of these methods can be shown either more or less rational than any of the others. But if every method is equally lacking in rational justification, then there is no method which can be said to furnish a rational basis for prediction, for any prediction will be just as unfounded rationally as any other.

Now replace 'prediction' with 'criticism' (and 'predictions' with 'criticisms') in the foregoing paragraph. I think this nicely illustrates my point:

> The question is not whether other methods—*e.g.*, astrology or numerology—provide more rational approaches to criticism than does the scientific method. The question is whether the scientific approach provides a more rational basis for criticism, for purposes of practical

action, than do these other methods. The position of the Humean skeptic would be, I should think, that none of these methods can be shown either more or less rational than any of the others. But if every method is equally lacking in rational justification, then there is no method which can be said to furnish a rational basis for criticism, for any criticism will be just as unfounded rationally as any other.

5.3. Method and Aim

True theories will give us true statements about the future (or accurate predictions); and likewise, true theories provide critical tools that allow us to rule out (only) practical proposals that will not succeed. (Again, the Duhem problem lurks in the background. For the moment, just assume that we can identify true auxiliaries, which we can conjoin with our theories for predictive and critical purposes, without difficulty.) But if corroboration is not linked to truth or even to verisimilitude, as suggested earlier, it appears reasonable to question why it should have anything to do with rational prediction. If corroboration were linked to empirical adequacy, at a bare minimum, that would suffice for solving the problem of rational prediction.

Now at some points, at least, Popper flirted with the notion that corroboration *is* an indicator of verisimilitude. This is clear from passages such as the following:

> If two competing theories have been criticized and tested as thoroughly as we could manage, with the result that the degree of corroboration of one of them is greater than that of the other, we will, in general, have *reason to believe* that the first is a better approximation to the truth than the second.[39] (Popper 1983, p. 58)

Yet it is hard to reconcile this with what Popper said, to my mind correctly, elsewhere:

> As to degree of corroboration, it is nothing but a measure of the degree to which a hypothesis *h* has been tested, and of the degree to which it has stood up to tests. (1959, p. 415)

> Corroboration (or degree of corroboration) is thus an evaluating *report of past performance* . . . Being a report of past performance only . . . *it says nothing whatever about future performance*. (1972, p. 18)

In fact, corroboration does not indicate truth or verisimilitude (if induction is not appealed to). Watkins (1984, pp. 284–285) provided the clearest and simplest explanation of why, which I summarise as follows:

(1) Corroboration says nothing about future performance.

(2) Verisimilitude can only be judged on the basis of future performance.

Therefore,

(3) Corroboration says nothing about verisimilitude.

As a consequence of their rejection of induction, (1) is accepted by critical rationalists. The fact that a theory has passed one test in the past does not increase the probability that it will survive a different test in the future (even when the existence of spatio-temporally invariant laws is assumed). And (2) is easy to see with even a simple hypothesis like "All bunnies are brown"; clearly we could only tell if this were true (and how true, if one allows such a notion) if we could see bunnies in the future. (3) follows.

In passages such as the following, Popper recognised this:

> From a rational point of view we should not 'rely' on any theory, for no theory has been shown to be true, or can be shown to be true . . . But we should *prefer* as basis for action the best-tested theory . . . the best-tested theory is the one which, in the light of our *critical discussion* appears to be the best so far . . . in spite of the 'rationality' of choosing the best-tested theory as a basis for action, this choice is *not* 'rational' in the sense that it is based upon *good reasons* for expecting that it will in practice be a successful choice: *there can be no good reasons in this sense, and this is precisely Hume's result.* (1972, pp. 21–22)

So, in short, corroboration *is* a measure of how hard we've tried to refute something, and how resilient it has proven so far. The aim in testing it is to show it to be false, if it is false; no more, and no less. Of course, one might ask, 'Why assume that trying hard to refute something results in refuting it if it is false?' First, it should be noted that if we assume the reliability of observation statements (which many of us are happy to do), then falsifications are also reliable (if Duhem's problem is put to one side); that's to say reliable observations of the colour of rabbits will lead to reliable refutations of false universal hypotheses about the colour of rabbits, and so forth. Second, we come to the argument promised in the previous subsection.

To see how more testing can increase the chance of identifying false theories, consider the following simple scenario. Imagine there are five possible tests of a theory, only one of which will show it to be false (e.g. because it is empirically inadequate). Picking a test at random gives a one in five chance of falsifying the theory. If that test fails to falsify the theory, then we may exclude it (and have four possible tests remaining). Picking another test at random will give a one in four chance of falsifying the theory, and the overall chance of having falsified the theory after two tests selected at random will be $\frac{2}{5}$, i.e. $1 - \frac{4}{5} * \frac{3}{4}$. And so on.

This may be generalized as follows: if there are n possible tests and m of those tests will identify T as false if performed, then the probability of selecting a test which falsifies T (at random) is (m/n). If we also specify that when a test is performed which fails to falsify T then it is not performed in future (at least until all other tests have been explored), then the probability of randomly selecting a test which falsifies T, after the completion of f tests which have failed to falsify T, is $(\frac{m}{n-f})$ where $m \geq 0$, $n \geq m$ and $n > f \geq 0$. Each time a test fails to falsify T, the chance of achieving falsification with the next test increases; that is, provided that m is not zero. The overall chance of achieving falsification steadily increases with each test.

Naturally this argument works on the basis, first, that only a finite number of tests are possible. And while this is false if 'possible' is understood to reflect what is possible in principle, it is plausibly true if it is understood to reflect what is possible in practice. Yes, it must be confessed that no two tests can be strictly identical because background conditions will always change; the state of the universe is never the same twice, and no system (except the universe itself) is ever truly closed. However, it does not follow that an infinite number of *significantly different* tests are possible in practice; and we may therefore consider 'tests' to refer to 'significantly different tests'. (We may be wrong about which differences are significant, of course, but this brings us on to the following point.)

Second, as noted earlier, none of the possible tests for any given false theory may be sufficient to falsify said theory (i.e. m may be zero). And in that event, repeated testing will not, of course, increase the chance of falsifying the theory. However, we are generally ignorant of whether we have at our disposal some test which will falsify any given theory (unless we are already satisfied that the theory is false). So while it would be remiss of us to blithely assume that we have no such test available, and fail to search for it (in the hope that it exists), all we truly *know* is that the chance of identifying some theory as false, if it is false, increases with the more tests we perform *provided that we are able to perform a test that can show it to be false.* (It could be added that we may equally find ourselves in a situation where we have a true theory; but that is no excuse not to test it, because we cannot tell it is true!)

There are also, naturally, other assumptions that have been made in this simple example: that we don't make mistakes in performing tests, that Duhem's problem doesn't prevent the falsification of theories, and so forth.[40] The point of the example, however, is to establish a basic principle. The fact that we make mistakes is a problem for the inductivist as much as it is for the critical rationalist; and Duhem's problem presents difficulties for confirmation theorists too. It is therefore unhelpful to complicate the discussion in the present context.

5.4. Why Trust Observations?

One might also wonder why we should trust observation statements on the critical rationalist account—why we should seek out 'basic statements' in order to falsify our theories, if they are no more secure than those theories themselves—and if induction will not need to be appealed to, ultimately, in order to make this case. It will not do merely to say that all falsification is conditional, 'asserting that *if* some test statements (potential falsifiers) are true, *then* a general theory is false' (Andersson 1994, p. 80), because conditional falsification can be done from the armchair. If there is a black rabbit, then "All rabbits are brown" is false. We can all see this. But what *motivates* empirical inquiry is presumably a hope (or belief) that we can work out whether this conditional falsification is of interest, i.e. whether there really is a black rabbit such as the one in Figure 2.3.

Figure 2.3 A photograph of Dr. Tim, a black rabbit.

However, it is not immediately obvious that appeal to induction would be appropriate to solve this problem. Even if we were to embrace the notion that scientists can and should use inductive methods, these only apply to propositions. Therefore a statement based on an experience cannot be justified by that experience *by inductive means*.

Nevertheless, one might say that we can construct some sort of argument for trusting our observations generally, e.g. by inference to the best explanation. What best explains the character and nature of our experiences, it might be said, is that there is a world of mind-independent objects that we perceive reasonably accurately most of the time, and so on. (Evolutionary arguments, concerning what best explains our survival, may also be employed.) To this, the critical rationalist will suggest that there is no evidence that our observation statements are generally reliable, and furthermore that they are *always* fallible. (Seriously, what is the evidence, say, that there is a table *qua* physical object in front of me? I have never heard a satisfactory answer to this question. Most people will say "You can touch it", "You can see it", and so forth. But that won't do, because crucial metaphysical premises are missing for which no evidence is provided. And appeal to lofty scientific theories such as evolution, which are corroborated only on the basis of observation statements themselves, if at all, will not do. Only circularity will result.) She will add that this should hardly be surprising, given the fundamentally theoretical and transcendent character of observation statements (which I mentioned, but chose to temporarily ignore for the purposes of argument, towards the beginning of the chapter). In the words of Popper (1981, p. 88), recall: 'All observations are theory impregnated'.

This is true even of the simplest statements of experience which are mistakenly said by Russell (1912) to be indubitable.[41] Keuth (2005, p. 100) makes this point clearly:

> Let us now assume that when I say "This area now looks red," I mean only that it appears red *to me* here and now. Have I thus finally eliminated *any* transcendence inherent in my description? By using the predicate "red," I presuppose that my present colour impression equals the impressions that I have had on other occasions when I have used the same word. Hence I try to use it *according to a rule*. However, I do not now have the other impressions; rather, I only remember them. Accordingly, I still assert more than I sense here and now. *As long as we are using predicates at all, we cannot avoid this kind of transcendence inherent in our descriptions*. But subject-predicate statements are the simplest kind of statements in our language. *For that reason alone, no perception can secure the truth of any statement*.[42]

Interestingly, a similar point was made independently of Popper, and somewhat earlier, by Morris Cohen (1953 [1931], pp. 124–125):

> That all knowledge begins with the perception of the individual and then goes on by abstraction to the universal is a widespread dogma . . . We are impressed with a stranger's beauty, agreeableness, or reliability before we can specify his features or traits. It is therefore quite in harmony with fact to urge that the perception of universals is as primary as the perception of particulars. The process of reflection is necessary to make the universal clear and distinct, but as the discriminating element in observation it aids us to recognize the individual . . . A student will make little progress in geometry if his attention is solicited by the special features of his particular diagram rather than by the universal relations which the diagram imperfectly embodies . . . without some perception of the abstract or universal traits which the new shares with the old, we cannot recognize or discover new truths.

So however unsatisfactory it may seem to some—and I return to this topic in the concluding chapter—the critical rationalist may suggest that the typical reliability of (a class of unproblematic) everyday observation statements (which are theory laden, often with traditional folk theories) is assumed on pragmatic grounds, to enable inquiry to proceed. Like metaphysical realism on Rescher's (1987, p. 126) account discussed previously, in other words, we may say this is a postulate accepted 'on *functional* rather than *evidential* grounds'. (And if inductivists such as Rescher accept that we need to do this for something as fundamental as metaphysical realism, then it is hard to see why we shouldn't do it in the present context. After all, the reliability of our observations does not follow from metaphysical realism.)

It is also crucial to recognise that observation statements—or what Andersson (1994) calls 'test statements', which ought to be assessable inter-subjectively, in preference to Popper's 'basic statements'—may typically be tested by other such statements:

> From a logical point of view, we are never forced to stop at a particular type of test statement. In actual research we use such test statements that can easily be tested inter-subjectively. Often auxiliary hypotheses are implicitly presupposed, and the test statements are theory impregnated and far away from "pure" observations.

In the words of Popper, which Andersson (1994, p. 79) also cites:

> [W]e are stopping at statements about whose acceptance or rejection the various investigators are likely to reach agreement. And if they do not agree, they will simply continue with the tests, or else start them all over again. If this too leads to no result, then we might say that the statements in question were not inter-subjectively testable, or that we were not, after all, dealing with observable events. (Popper 1959, p. 104)

Sticking one's head in the ground is an option instead, of course; the only rational one, to boot, if one should never believe beyond one's evidence. Critical rationalists are rather more optimistic in the face of the considerable difficulties that they believe we face. Yes, the assumption that our observation statements are true more often than false may be wrong. But what then should we do? Admit defeat, presumably.

Inductivists trust observation statements—subject to similar provisos as those introduced by critical rationalists, about intersubjective testability, and so forth—too. The critical rationalist doubts that there are evidential reasons for doing so. This is not to concede that science (or inquiry more generally) is irrational, because to deny the presence of evidential reasons is not to deny the presence of reasons.

5.5. An Inductive View of the Value of Severe Testing?

Some would claim there is an *inductive* way to capture the insight that evidence is important primarily because of its potential to correct error; or in other words, that one may have an inductive account of the significance of testing. Mayo and Spanos (2006, p. 328), in their discussion of the Neyman-Pearson (N-P) tests, argue that:

> N–P tests can (and often do) supply tools for inductive inference by providing methods for evaluating the severity or probativeness of tests. An inductive inference, in this conception, takes the form of inferring hypotheses or claims that survive severe tests. In the 'severe testing' philosophy of induction, the quantitative assessment offered by error probabilities tells us not 'how probable', but rather, 'how well probed' hypotheses are.

In order to evaluate this claim, it may be helpful first to note the way in which Neyman (1957) construes N-P tests (which were introduced in Neyman and Pearson 1933) as rules of inductive behaviour. Mayo and Spanos (2006, p. 326) explain the basic idea as follows:

> Why should one accept/reject statistical hypotheses in accordance with a test rule with good error probabilities? The inductive behaviorist has a ready answer:
>
> > Behavioristic rationale: We are justified in 'accepting/rejecting' hypotheses in accordance with tests having low error probabilities because we will rarely err in repeated applications.

However, the same sort of behaviouristic rationale may be appropriate even beyond the context of statistical hypotheses. That is to say, it may be suggested that if we only ever accept *non-statistical* hypotheses when the probability that they are incorrect is low, then we will rarely err. Yet it is crucial to see that this is only if the sort of probability under consideration is aleatory (i.e. frequency or propensity based) rather than epistemic (e.g. subjective or logical). What we want, that is to say, is some reliable process by which to rule out false (or falser) theories, or rule in true (or truer) theories. We discussed something similar earlier in this chapter.

The basic move of Mayo and Spanos (2006, p. 330) is to shift the emphasis from the long run to individual cases. They offer the following nice example:

> Suppose a student has scored very high on a challenging test—that is, she earns a score that accords well with a student who has mastered the material. Suppose further that it would be extraordinary for a student who had not mastered most of the material to have scored as high, or higher than, she did. What warrants inferring that this score is good evidence that she has mastered most of the material? The behavioristic rationale would be that to always infer a student's mastery of the material just when they scored this high, or higher, would rarely be wrong in the long run. The severity rationale, by contrast, would be that this inference is warranted because of what the high score indicates about this student—mastery of the material.

But we may ask how we are to know when we have tests that have low error probabilities. In some cases, this will admittedly be a simple *a priori* (or definitional) matter; consider, for example, testing whether a coin is fair by way of thousands of flips. But in many others, such a claim will be synthetic and far from evident. In general, it is not obvious how likely one is to discover that a theory is false (when it is false) by deriving the most unexpected predictions one can from it, and then looking for them. (And similarly, it is difficult to see how likely one is 'to discover' that a theory is false when it is actually true by adopting such a procedure.) Moreover, if we are to know that there is such a reliable process, we will have to rely on our personal (or intersubjective) evidence, and then we will be back to worrying about subjective approaches to confirmation.

Nevertheless, it is interesting to note that *methodologically speaking*, Mayo and Spanos's (2006) recommendation appears to be barely distinguishable from Popper's. Grant for the sake of argument that we all agree on how to measure the severity of tests, although doing this is not as easy as it may first seem (as I argue in Chapter 4). Broadly, Popper says we should prefer highly corroborated theories, when it comes to some theoretical and practical contexts at least, to their less corroborated (or uncorroborated) counterparts. Mayo and Spanos (2006, p. 328) appear to differ only in so far as they suggest that we should *infer* (rather than simply *prefer*) 'hypotheses or claims that survive severe tests'. They say the difference stems from the fact that: 'Popper, and the modern day "critical rationalists" deny they are commending a reliable process' (ibid.).

This is not the right way to summarise the difference, though. Imagine we entertain the hypothesis that we *know for sure* that some procedure is 99 per cent reliable. Let's say that we have some machine which is designed to select black balls from a bag of balls of mixed colours, but which only has a 0.99 propensity—we miraculously know beyond any possible doubt—of succeeding in any given case.[43] Popper, like any other critical rationalist (to the best of my knowledge), would gladly accept, in such a scenario, that it would be unreasonable not to expect the ball to be black. In fact, we can go a step further by imagining we were forced into a bet for a marginal stake S, against an opponent who (we know) may freely choose whether she bets for or against a 'black' result, and asked to select a betting quotient q on the understanding that we will either (A) pay qS and receive S on a 'black' result (in the event that our opponent bets against 'black') or (B) be paid qS and have to pay out S on a 'black' result (in the event that our opponent bets on black). Now we may contently advocate the view that the correct choice of betting quotient (q) is 0.99 without making any concessions on the issue of induction. (The probability in this example is a guide to action because it is *aleatory* rather than epistemic, i.e. reflects a genuine chance in the world. Critical rationalists object to the view that we should make inductive inferences, recall, precisely because they reject the view that high inductive probability, as it is typically defined, can be interpreted in an aleatory fashion.) Admittedly, the example grants the existence of what Popper doubts, namely, *knowing for sure* about (the degree of) the reliability of some process. But Popper does not doubt that reliable processes are possible, or indeed that we can ever (fallibly) identify them.[44]

5.6. The Argument from Artificial Intelligence

Last but not least, we should consider the argument, presented in the philosophical literature primarily by Gillies (1996, 2003), that mechanical induction has been shown to be possible by advances in artificial intelligence.[45] Mechanical induction involves simply 'a mechanical method of obtaining scientific laws from a large mass of data previously collected' (Gillies 1996, p. 3). And if this does indeed work, then it would show that Popper (1983, p. 6) was incorrect to declare that: 'There is no method of discovering a scientific theory . . . There is no method of ascertaining whether a hypothesis is "probable", or probably true.'

It is worth adding, because Carnap and Popper are so often painted as being entirely at odds on the issue of induction, that Carnap (1962, pp. 192–193) actually *agreed* with Popper about the impossibility of mechanical induction (while nonetheless thinking that confirmation, above and beyond corroboration, *is* possible):

> [T]he inductive procedure is not, so to speak, a mechanical procedure prescribed by fixed rules. If, for instance, a report of observational

> results is given, and we want to find a hypothesis which is well confirmed and furnishes a good explanation for the events observed, then there is no set of fixed rules which would lead us automatically to the best hypothesis or even a good one. It is a matter of ingenuity and luck for the scientist to hit upon a suitable hypothesis . . . The same point has sometimes been formulated by saying that it is not possible to construct an inductive machine. The latter is presumably meant as a mechanical contrivance which, when fed an observational report, would furnish a suitable hypothesis, just as a computing machine when supplied with two factors furnishes their product . . . [A]n inductive machine of *this* kind is not possible.[46]

Nonetheless, Gillies (2003) claims that:

> Advances in artificial intelligence have, however, shown that they [i.e. Carnap and Popper] were both wrong on this point. In fact programs have been written which enable computers, when fed with data, to generate suitable hypotheses for explaining that data. Moreover this new kind of computer induction has resulted in the discovery of important and previously unknown scientific laws.

We should note from the start, however, that the programmes examined by Gillies (1996) involve a process of iteration in order to arrive at their final results, and that 'testing and falsification always play a key role in this iteration process' (Gillies 1996, p. 18). As such, it should be noted that mechanical induction (of the form allegedly performed by AI) may involve what was earlier called an inductive-deductive process. A body of data is collected. A theory is induced. It is tested. If it fails, a replacement theory is then induced. And so on. (We will come back to discuss whether such processes are liable to lead us to the truth, or even just empirical adequacy, in the discussion of evolutionary epistemology in Chapter 7.)

This said, it seems to me that the empirical possibility of a refutation of Popper's (and the critical rationalist's) stance on mechanical induction, and therefore on the legitimacy of inductive inferences (as a part of that process), is one that we should take seriously. Just as we would accept that a die was fair if the frequency of each number was approximately the same in a large number of rolls of the die, so we should accept that mechanical induction is possible if artificial intelligences can display repeated consistent successes in deriving laws of nature by applying inductive rules for hypothesis formation (and testing procedures).[47] And if inductive inferences are a part of that process, then we should accept their importance.

Now without entering into unnecessary detail, let us look at what Gillies (1996, p. 53; 2003) takes to be one of the best examples of a law derived by an AI (namely GOLEM) considering the secondary structure of proteins:

GOLEM'S Rule 12 regarding Protein Secondary Structure

There is an α-helix residue in protein A at position B if
(i) the residue at B-2 is not proline,
(ii) the residue at B-1 is neither aromatic nor proline,
(iii) the residue at B is large, not aromatic, and not lysine,
(iv) the residue at B+1 is hydrophobic and not lysine,
(v) the residue at B+2 is neither aromatic nor proline,
(vi) the residue at B+3 is neither aromatic nor proline, and either small or polar, and
(vii) the residue at B+4 is hydrophobic and not lysine.[48]

Gillies (1996, p. 53; 2003) then adds:

> Some readers may feel rather disappointed with this rule, which is rather long, cumbersome, and specific. It was, however, 95% accurate on the training set, and 81% accurate on the test set. It was not known before being produced by GOLEM, and it makes a contribution to an important current problem in the natural sciences. It seems to me fair, therefore, to credit GOLEM with the discovery of a law of nature.

I do not think we should object that the rule is long, cumbersome, and specific. On the contrary, it has the correct form of a law-like statement. It is a statement about *all* proteins which satisfy the conditions (i)–(vii) (and therefore possess the properties specified therein); and it says that they will have a particular further property.

Yet I do not think it is a law, since the statement is false! To this, I presume the response will be that the proposed law was *statistical* in nature. But I do not think it was, because the prediction did not have a probabilistic form. That is to say, the statement of the rule did not take the form "There is an alpha helix residue in protein A at position B, *with probability P*, if . . ." And note that even if were we to read the result as '*with probability P greater than 0.5*', the statement still does not have a law-like form, e.g. on the account discussed by Hempel (1965) in his treatment of the inductive-statistical form of scientific explanation.

It is as if we had a programme derive the result "All rabbits are brown" after examining a group of rabbits with regard to several different properties. Imagine we then used a very large test group of bunnies, previously unexamined, and discovered that 81 per cent of them were brown. Would we credit the programme with a great discovery? Hardly! The programme would not have told us that rabbits have a probability of 0.81 (or something in the region) of being brown. It would have derived a false statement.

Yes, it is easy for us to spot that a lot of rabbits are brown although it is difficult for us to spot that a lot of proteins meeting conditions (i)–(vii) also have alpha helix residues at position B. And a computer programme may help us to spot correlations such as this, in groups of things, which we may otherwise miss. But to do this is one thing. To arrive at a *law*, statistical or otherwise, is quite another.

In the case of GOLEM's discovery, moreover, the question then remains as to whether the correlation requires a scientific explanation *in terms of fundamental laws* at all, and what they are (if so). (Some patterns need explaining, whereas others don't. A programme like GOLEM cannot tell these apart.) In short, one might also object that the rule proposed by GOLEM is merely phenomenological (even if it is non-accidental). That is to say, the link spotted may well be *due to* fundamental physical or chemical laws, but is not itself such a law. Whether this is a good objection is a matter of some controversy; for Cartwright (1983), for instance, phenomenological laws are more 'real' than their fundamental counterparts (because of all the idealizations necessary to bring the latter into contact with experience in an experimental context). But even if GOLEM's rule were a possible statistical law, which it is not, then it would certainly not be a fundamental one.

This said, critical rationalists should be open to, and thankful for, the possibility of an empirical refutation of their views on induction in the future.[49] As Gillies was right to emphasise in his correspondence with me, computing power is increasing all the time.

ACKNOWLEDGEMENTS

Section 3 of this chapter is based on Rowbottom (2006a).

3 Corroboration and the Interpretation of Probability

A standard criticism of Popper's philosophy, summed up neatly by Curd and Cover (1998, p. 508), is: 'it is hard to see how [he] can justifiably assert that science is rational and objectively progressive when it ultimately depends on purely conventional and arbitrary decisions'. Typically, critics have focused on his discussion of 'The Empirical Basis', in which he writes (Popper 1959, pp. 108–109):

> From a logical point of view, the testing of a theory depends upon basic statements whose acceptance and rejection, in its turn, depends upon our decisions. Thus it is decisions which settle the fate of theories. To this extent my answer to the question, 'how do we select a theory?' resembles that given by the conventionalist . . .

Yet while this debate, to which I offered a pragmatic resolution in the previous chapter, is well worn—see Ayer (1974), O'Hear (1980, ch. 5), Newton-Smith (1981, pp. 59–64), Bartley (1984, app. 3), Haack (1991), Miller (1994, pp. 29–30), Andersson (1994), and Zahar (1995)—the same fundamental objection might be arrived at by another immanent critique, which hinges on the issue of how epistemic probabilities are to be interpreted.

As we will see, the problem lies with the logical interpretation of probability favoured by Popper in epistemic contexts.[1] But if we instead admit that subjective probabilities are better suited to the corroboration function than logical ones, what counts as 'corroborated' appears to be a purely psychological matter. This criticism leaves us in rather a bind. However, there is a now an alternative epistemic interpretation of probability which we might consider in place of the subjective one: the intersubjective interpretation of Gillies (1991, 2000).

After illustrating the problem(s) with the logical interpretation of probability and with adopting a subjective view of probability for measuring corroboration, my goal in this chapter is to show that intersubjective probability assignments are generally superior to subjective ones, when it comes to scientific decision making (*inter alia*), given a number of plausible constraints. I will then argue that intersubjective corroboration is preferable to intersubjective Bayesian confirmation in a key respect, namely that agreement on corroboration values—or narrow ranges thereof—is easier to achieve on principled, value-independent grounds.

So to summarise my argument in advance:

(1) There are only two acceptable epistemic interpretations of probability: (a) subjective and (b) intersubjective.
(2) Intersubjective probabilities are superior for the purposes of determining Bayesian confirmation or corroboration.
(3) Intersubjective Bayesian confirmation values cannot be determined in a value-free way and agreement on said confirmation values is therefore difficult and of dubious epistemic significance.
(4) Intersubjective corroboration values can be determined in a value-free way.
(5) Intersubjective corroboration is therefore a superior measure to intersubjective Bayesian confirmation.

1. CORROBORATION AND THE LOGICAL INTERPRETATION OF PROBABILITY

Recall the corroboration function, which measures the corroboration of a hypothesis *h*, given evidence *e* and background knowledge *b*:

$$C(h,e,b) = \frac{P(e,hb) - P(e,b)}{P(e,hb) - P(eh,b) + P(e,b)}$$

When I introduced this in the previous chapter, I used the example of the Poisson (or Arago) bright spot, which was really a rediscovery of a phenomenon earlier noted by Maraldi (Hecht 1998, p. 486). Let's now think about how we should interpret the probabilities in the function, with reference back to that example.

We can all agree that the experiment showing the bright spot impressed several scientists, and that it had a profound psychological effect on many who witnessed it or heard about it. But what objective significance did it have, if any? The answer seems to be none, beyond the trivial consequence that a new phenomenon was catalogued, if our understanding of epistemic probability is subjectivist—if we take probabilities to reflect synchronically coherent degrees of belief—just because what seems 'impressive' is a rather arbitrary matter.[2] If a new theory *T* is employed to make a successful prediction that you presently think is incredibly unlikely, you might be shocked. You might then be psychologically predisposed to believe *T*. But would this be a *reason* to prefer it to its competitors, even if those had been used to make many more successful, but just less surprising, predictions?[3] It seems clear that the answer lies in the negative.

Corroboration values are significant in an objective sense, for Popper, because he subscribes to a logical interpretation of probability: because he believes in logical relations between propositions (or groups thereof) other than entailment. So the probability of Newton's theory of gravitation relative to a given set of propositions (e.g. observation statements) has a single, definite, immutable value.[4] It does not have a different value for me than it does for you, which would be reflected in the different odds we would accept for a bet on the truth of the theory, given the same background knowledge. As we saw in the previous chapter, indeed, Popper defended the view that the probability of any universal law is zero. He did not take this to be a statement of his personal (coherent) opinion!

At this stage some words of warning are in order, since there has been considerable confusion about the role of corroboration in Popper's philosophy, due, for instance, to Putnam (1969) as explained by Popper (1974d). First, Popper (1959, p. 418) states a specific domain in which his function gives a meaningful value:

> C(*h*,*e*) can be interpreted as a degree of corroboration only if *e* is a report on the severest tests we have been able to design . . . [and] C(*h*,*e*) must not be interpreted as the degree of corroboration of *h* by *e*, unless *e* reports the results of our sincere efforts to overthrow *h*.

We will come back to this in the next chapter; but for the moment, note that accidentally/incidentally acquired information, e.g. casual observations (or more properly statements made on the basis of said observations), cannot ever corroborate a theory on this view.

Second, Popper (1983, pp. 254–255) does not suggest that scientists should calculate corroboration values (except perhaps, as we will see, when statistical laws are under consideration):

> I do not believe that my definition of degree of corroboration is a contribution to science except, perhaps, that it may be useful as an appraisal of statistical tests . . . Nor do I believe that it makes a positive contribution to methodology or to philosophy—except in the sense that it may help (or so I hope) to clear up the great confusions which have sprung from the prejudice of induction, and from the prejudice that we aim in science at high probabilities—in the sense of the probability calculus—rather than at high content, and at severe tests.

This said, we can get back to the issue at hand. Even if scientists need usually not spend their time calculating corroboration values, we still need to be able to show that such values genuinely exist (and can be roughly determined, without rigorous calculation, at least). In the words of De Finetti (1972, p. 23): 'For any proposed interpretation of Probability, a proper operational definition must be worked out: that is, a device apt to measure it must be constructed.'[5] So how is one to measure probabilities on a logical account like Popper's?

Keynes (1921, p. 41), the architect of the logical interpretation, suggested, 'In order that numerical measurement may be possible, we must be given a number of equally probable alternatives'.[6] But how are we to determine when we are faced with a number of equipossible alternatives?[7] Enter the principle of indifference, the *a priori* synthetic principle that was previously dubbed 'The Principle of Non-Sufficient Reason' by Bernoulli:

> The Principle of Indifference asserts that if there is no *known* reason for predicating of our subject one rather than another of several alternatives, then relatively to such knowledge the assertions of each of these alternatives have an *equal* probability. Thus *equal* probabilities must be assigned to each of several arguments, if there is an absence of positive ground for assigning *unequal* ones. (Keynes 1921, p. 42)

As Keynes (1921, p. 41) himself admits, however, in application this principle 'may lead to paradoxical and even contradictory conclusions'. There are several examples of such paradoxes presented in Gillies (2000), one of which is the geometrical paradox of Bertrand (1960 [1889], p. 4): 'Draw a *random* chord in a circle. What is the probability that it is shorter than the side of the inscribed equilateral triangle?' (translation mine).[8]

So let us consider an equilateral triangle with centre O, inscribed in a circle with radius R:

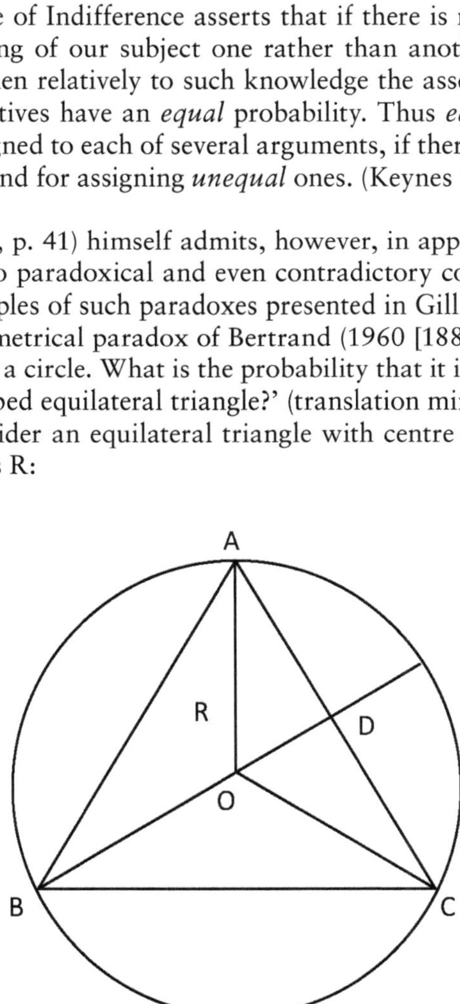

Figure 3.1 Depiction of Bertrand's paradox.

The chord passing through B and O—a diameter—bisects AC. Thus, <ODC> is a right angle, and OD = Rsin30 = R/2. Now our problem: if we select a chord of the circle at random, what is the probability that it will have a length greater than the side of the triangle ABC, $P(\Psi)$? First, we might let XY be a random chord and OZ be the line bisecting XY at point W:

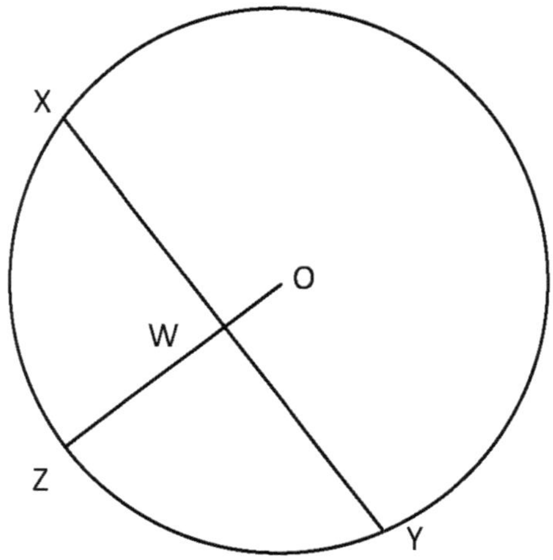

Figure 3.2 Solution One.

Now we have no known reason to presume that W is at any particular point on OZ rather than any other, thus all points along OZ are equipossible locations thereof by the principle of indifference. In other words, OW has a uniform probability density in the interval [0, R]. And XY will be longer than the side of the triangle ABC if and only if OW is less than R/2. Thus,

$$P(\Psi) = P(OW < R/2) = ½ \qquad \text{(Result 1)}$$

Second, let AA' be a chord of the circle, with an angle θ to the tangent to the circle at point A, as depicted here:

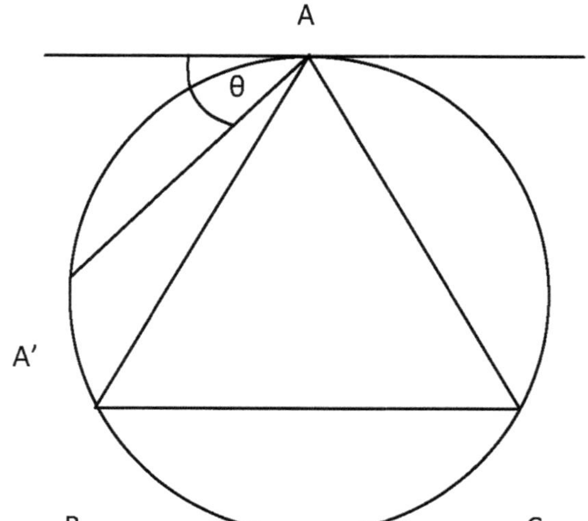

Figure 3.3 Solution Two.

If AA' is to be longer than the side of the triangle, θ must be between π/3 radians and 2π/3 radians. And we have no known reason to suppose that θ has any particular value between 0 and π radians, rather than any other, hence by the principle of indifference θ has a uniform probability density in the interval [0, π]. Thus,

$$P(\Psi) = P(\pi/3 < \theta < 2\pi/3) = \tfrac{1}{3} \qquad \text{(Result 2)}$$

Third, and finally, let us inscribe a circle in the triangle ABC—the 'secondary circle'—with a radius of R/2. And let the chord be XY, drawn between any two distinct points on the primary circle's circumference. The situation is then as depicted in Figure 3.4 (with the triangle ABC omitted, for clarity):

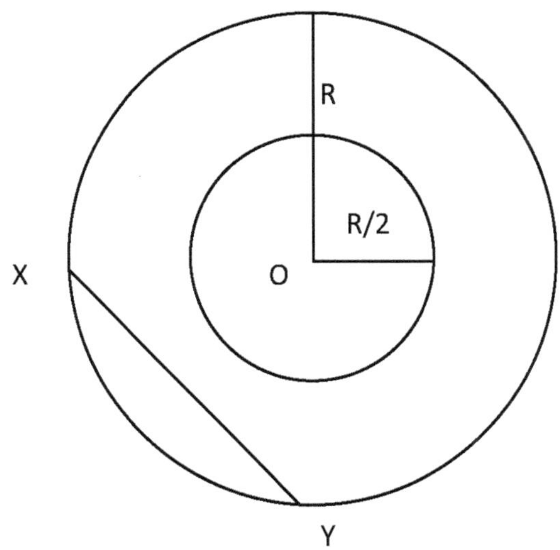

Figure 3.4 Solution Three.

If XY is to be longer than the side of ABC, then its central point—call this W—must lie inside the secondary circle. And we have no known reason to presume that W lies at any point inside the primary circle rather than any other, thus W has a uniform probability density in the primary circle according to the principle of indifference. It follows that:

$$P(\Psi) = \text{Area of secondary circle/Area of primary circle} = \pi R^2/4\pi R^2 = \tfrac{1}{4} \qquad \text{(Result 3)}$$

This third case should be excluded, however, because Bertrand failed to notice that there are infinitely many chords that have their centre as O—that is, the centre of the circles—but for any other point inside the primary circle, there is only one chord with that point as its central point. Perhaps one could, however, use this result to argue that the probability of choosing a central chord is the same as the probability of choosing a chord elsewhere, and thereby arrive at a suitable variant of the third version. No matter; the conflict between results 1 and 2 is sufficient to generate a paradox.

Bertrand (1960 [1889], p. 5) concludes: 'The question is badly posed.' Jaynes (1973), however, provides an argument that just one of these candidate answers, namely 1, is correct. Jaynes's idea is that the correct solution must obey invariance principles, with respect to rotation, scale, and translation. He also claims to have tested (and corroborated) result 1 empirically, after selecting it on the basis of said principles, on the (highly dubious) assumption that it is possible to perform an empirical test without unacceptable idealizations.

But even if Jaynes is right that invariance principles can solve certain apparent paradoxes, it remains the case that there might have been 'no known reason' for employing them in the earlier example, on the part of a hypothetical theorist, and the results obtained would, in fact, have been contradictory. And it is not acceptable to just apply the principle, see that it fails in a particular case, and then say "There is something we should have known, or realized, or that we knew without knowing that we knew". For that would be a means of immunising the principle, *qua* alleged *a priori* and synthetic truth, from criticism. Therefore the principle must be altered in so far as it must be 'topped up' by a requirement such as "use natural measures", at the very least. But in one of the most recent papers on Bertrand's paradox, Shackel (2007, p. 174) argues that:

> There is, so far as we can know, no "natural" measure on the set of chords. No one has succeeded in justifying the claim that any particular measure on the set of chords is the correct measure for the general problem. All in all, then, there is no reason to think that the well-posing strategy can succeed.

Beyond the narrow confines of Bertrand's paradox, indeed, it is exceedingly difficult to see how all possible paradoxes could be evaded by appeal to an overarching scheme of 'natural' classification (Gillies 2000, pp. 41–42):

> It is easy to see how we can generalise . . . to produce a paradox in any case which concerns a continuous parameter (θ say) which takes values in an interval $[a, b]$. All we have to do is consider $\varphi = f(\theta)$, where f is a continuous and suitably regular function in the interval $[a, b]$ so that $a \leq \theta \leq b$ is logically equivalent to $f(a) \leq \varphi \leq f(b)$. If we have no reason to suppose that θ is at one point of the interval $[a, b]$ rather than another, we can then use the Principle of Indifference to give θ a uniform probability density in $[a, b]$. However, we have correspondingly no reason to suppose that φ is at one point of the interval $[f(a), f(\mathrm{b})]$ rather than another. So it seems we can equally well use the Principle of Indifference to give φ a uniform probability density in $[f(a), f(\mathrm{b})]$. However, the

> probabilities based on θ having a uniform probability density will in general be different from those based on φ having a uniform probability density; and thus the Principle of Indifference leads to contradictions.[9]

Popper expresses himself somewhat differently from Keynes, in saying that P(*a*,*b*) measures the logical proximity of *a* to *b* or 'the degree to which the statement *a* contains information which is contained by *b*' (Popper 1983, p. 292). He does agree with Keynes that conditional probabilities are fundamental, as Hájek (2003) also argues, though. Even when it comes to his formal account of probability, rather than any specific interpretation thereof, Popper prefers to define absolute probabilities in terms of conditional ones (see Popper 1959, p. 321, fn. *1 and app. *iv). But he nevertheless gives the following example of a specific numerical relation:

> The 'logical interpretation' takes the probability calculus as a generalisation of ordinary logic, as it were . . .
>
> The intuitive justification runs as follows. Let *a* be the statement 'Socrates is mortal' and *b* be the statement 'All men are mortal and Socrates is a man'; then we shall say that p(*a*,*b*) = 1, because *a* follows from *b*; and indeed, given *b*, we may consider *a* as certain . . . But let *a* be the same statement as before, and *b* the statement '92 per cent of all men are mortal, and Socrates is a man', then *a* will not be certain on the information *b*, but highly probable; and we may indeed say that the probability which on the information *b* is attached to *a* will not be far from 0.92; that is to say, p(*a*,*b*) will be about 0.92. (Popper 1983, p. 293)

As intuitive as this justification might be, it is plausibly flawed. Indeed, the suggestion that P(*a*,*b*) will only be *about* 0.92 looks like a significant concession. The charge is rather simple: there is a veiled use of the principle of indifference, or something similar, here. The line of reasoning might begin with 'We have no reason to suppose that Socrates is any particular man (in the class of men) rather than any other'. It would end with a serious mistake.

In fact, it would seem that the information is radically insufficient to say anything about the probability which ought to be assigned to *a* on the basis of *b*, let alone the objective proximity of *a* to *b*. In particular, if '92 per cent of all men are mortal' is taken to express a relative frequency, what licenses the jump to any numerical assignment of the probability of Socrates, *qua* individual member of the class of men, being mortal? We can imagine adding a proposition *c*, 'Socrates is as likely to be any one man as any other', which would seem to yield P(*a*,*bc*) = 0.92, but then we can equally imagine adding a proposition *d*, 'Socrates is necessarily immortal', which would seem to yield p(*a*,*bd*) = 0. And no matter how intuitively appealing the leap to P(*a*,*b*) = 0.92 might seem, it would also seem to be a case in which our intuition leads us astray.

The point might be pressed further by the recognition that there are ambiguities (both semantic and syntactic) in so far as statements, sentence types, or sentence tokens are concerned, with respect to the propositions that they are intended to express; and the use of statements as primary truth bearers therefore seems rather curious, at least if the arguments of Alston (1996, ch. 1), from the point of view of the correspondence theory of truth which Popper also favours, are right. In short, the statement that '92 per cent of all men are mortal' might be understood to pick out a proposition about the propensity of each individual man to be mortal, in which case the conclusion $P(a,b) = 0.92$ is entailed by b.[10] As such it would be unobjectionable, but uninteresting with respect to any supposed 'degree of proximity' between *a* and *b*. And what is the probability that the statement '92 per cent of all men are mortal' refers to a propensity, rather than a relative frequency? What is it supposed to be conditioned on? While we have some notion of the closeness of two sentence types, with respect to what they could possibly specify in a given context (viz. if expressed as tokens of those types in a specific setting), this is linguistic (and conventional) rather than logical.

This leads to the conclusion that the logical interpretation of probability is no better off in Popper's variant than it is in Keynes's (or Carnap's). Even if we consent to the possibility of the existence of logical proximities, moreover, we might think that there are stronger critical reasons for believing in degrees of belief, and therefore prefer to use these when interpreting probabilities in epistemic contexts. (If a person can't be Dutch Booked, then her betting quotients must obey the axioms of probability. Assume that said quotients reflect rational degrees of belief, and you have some sort of case for the legitimacy of the subjective view. Stronger arguments have been attempted, as explained by Gillies [2000, pp. 59–62], but I take these to have defects. See Hájek [2005] and Rowbottom [2007b].) As if this were not enough, the description of Arago's discovery of the bright spot as merely 'very surprising to his fellow scientists' seems sufficient to the task of explaining why this was taken to be a great success for Fresnel's theory of light!

Admittedly, my discussion is incomplete to the extent that I have not considered new-fangled Objective Bayesianism, as championed by Jaynes (2003) and Williamson (2005, In Press). The basic idea behind this view is that we should start with the subjective view of probability, but then add on extra rationality constraints (in addition to coherence). Technically, one might argue that this is not a logical interpretation of probability; but how close it is may be seen by the fact that Keynes might equally have accepted the subjective interpretation and just 'tacked on' the principle of indifference as a rationality constraint, and then defined logical relations in exactly that way (rather than holding that those relations are 'out there' for us to intuit).[11] So as I have argued elsewhere—see Rowbottom (2008e)—the only radically new part of Objective Bayesianism, aside from issues in the philosophy of logic, is that a 'maximum entropy principle' replaces the principle of indifference. Jaynes (1957) introduced this principle long before he tried to provide a resolution to Bertrand's paradox, however, and I have already mentioned that Shackel (2007) shows that this attempt failed. Shackel and Rowbottom (In Progress) shows in detail how we should therefore reject Objective Bayesianism on the same grounds that we reject the logical interpretation.

2. CORROBORATION AND PROPENSITIES

Before we continue, we should briefly consider whether statistical laws should be exempted from the following discussion. The idea would be that when we are considering such laws we may understand the probabilities in the corroboration function as aleatory, e.g. as propensities. Imagine, for instance, that our hypothesis *h* is that some coin is fair (which we may interpret, more strictly, as meaning that coin flips under some specified repeatable conditions have a propensity of one half of landing on heads, and one half of landing on tails).[12] Now imagine that we decide to test this hypothesis, in the first instance, by flipping the coin one hundred times. Our evidence, *e*, is that the coin lands on tails every single time.

In order to evaluate C(*h*,*e*,*b*), we need to work out P(*e*,*hb*). And this is well defined when considered as a propensity in its own right, namely, as the propensity of that sort of experiment (i.e. any one hundred sequential flips under the specified repeatable conditions) to issue in *e*.[13] In general, if the chance of a 'tails' result is p and the result on each flip is independent of the other results, the probability of getting m 'tails' in n trials is:

$$P(m\text{ 'tails'}) = \frac{n!}{(n-m!)m!}\, p^m(1-p)^{n-m}$$

So if p is 0.5, then the probability of no heads results in n trials is 0.5^n. The probability of no heads in one hundred trials is therefore tiny—specifically, $7.89*10^{-31}$—and C(*h*,*e*,*b*) will be low no matter what the value of P(*e*,*b*).

Interpretative problems do arise, however, if we strive to understand P(*e*,*b*) as a propensity. It is easy to see this if we consider that the non-existence of propensities—and therefore the falsity not only of *h*, but also of the family of similar hypotheses where p has all possible values—might be consistent with *b*, and indeed that there may be no propensity whatsoever for *e* to occur in those circumstances. At first sight, setting P(*e*,*b*) to zero might seem to be an acceptable solution. There appears to be something wrong about doing so, however, because *e* is hardly ruled out, given realistic specifications of *b* (of the sort that many of us have), even if there is no propensity involving it.

The natural way to solve this problem is to introduce a methodological rule such that $P(e,hb)$ *qua* epistemic probability should match $P(e,hb)$ *qua* propensity in the event that h is a statistical law (i.e. a law that concerns propensities). We are then free to evaluate $C(h,e,b)$ in the foregoing scenario while interpreting it as a purely epistemic function (albeit one involving a hypothesis that concerns propensities in the world). We may now return to our consideration of epistemic probability.

3. SUBJECTIVISM VS. INTERSUBJECTIVISM

If the logical interpretation is replaced by the subjective interpretation, then it is unclear how there could be any deep objective rationality to science, beyond that of mob psychology. De Finetti (1937, p. 152)—an architect of the subjective interpretation, recall—writes:

> Our point of view remains in all cases the same: to show that there are rather profound psychological reasons which make the exact or approximate agreement that is observed between the opinions of different individuals very natural, but that there are no reasons, rational, positive, or metaphysical, that can give this fact any meaning beyond that of a simple agreement of subjective opinions.

Here's a simple example. Imagine I think that if I become a professor of philosophy, then this is a strong indication that God exists. Let p be 'God exists', and q be 'I become a professor of philosophy'; and let D denote a degree of belief and P a (subjective) probability. For me, $D(p,q) = 0.999$, which means that if I come to believe q then I'll be highly confident about p. For this also to be a probability, it need only be coherent (i.e. satisfy, along with related degrees of belief, the axioms of probability). So provided $D(\sim p,q) = 0.001$, and so forth, my degree of belief is a rational one; $D(p,q) = P(p,q) = 0.999$.[14] For one of my rivals, however, q might strongly indicate that p is false; and for her, $P(p,q)$ would be 0.001. What's more, crucially, we would both be rational *even if we shared precisely the same background information.*[15]

However, might there not be a middle ground between logical probabilities and personal degrees of belief? This is the idea presented by Gillies (1991; 2000, ch. 8). Intersubjective probability involves the extension of subjective probability from individuals to groups, and the central idea is that a group will do better if it agrees on a common betting quotient: 'that solidarity within a group protects it against an outside enemy' (Gillies 2000, p. 172). Furthermore, Gillies (1991, p. 519, fn. 2) tentatively suggests that nature herself can be considered an enemy, at least in so far as she can shatter our expectations, 'outwit' our best laid plans, and 'defeat' our most ingenious theories.

There is an obvious problem with this analogy, namely that nature works neither with us nor against us: it will not try to take advantage of our probability assignments. However, Gillies (1991, pp. 530–532) answers this objection by pointing out that the relevant game can be thought of as one played against the ideal experimenter (or experimenters). And this is a reasonable point: we wish to be as well prepared as we can be, given our human limitations.

Recall that for Popper an assessment of the corroboration of a theory has to be based on the severest tests available. But who is to determine what these are, if not a community of scientists? And wouldn't it be preferable if this decision were arrived at by critical discussion, in order to allow for correction of individual error? In fact, Popper (1983, p. 87) strongly emphasises the significance of interpersonal exchanges: 'We move, from the very start, in the field of intersubjectivity, of the give-and-take of proposals and of rational criticism.'[16] We might therefore suggest that intersubjective probabilities ought to be arrived at on the basis of appropriate critical activity. Let's now consider whether this is right.

Gillies (1991) shows that each member of a group ought to adopt the same degrees of belief (or betting quotients)—at least where shared interests are concerned—in order to prevent a Dutch Book being made against the collective. It's easy to see the basic principle by considering a simple scenario. Imagine a married couple who are avid gamblers with pooled financial resources. Romeo bets £100 that it will rain tomorrow in Oxford, at even odds. But unbeknownst to Romeo, Juliet has already bet £150, at three to one on, that it will not rain tomorrow in Oxford. (So Romeo bets as if the probability of rain is 0.5 although Juliet has already bet as if the probability of rain is only 0.25.) The couple is in trouble! They are 'out' £250. But no matter whether it rains in Oxford or not tomorrow, they will only get £200 back. As Hacking (1975, p. 8) points out, the possibility of such scenarios has been recognised since at least A.D. 9.

Clearly, Romeo and Juliet should have consulted each other and agreed on a betting strategy. But how, if at all, does this go any way to showing that such agreement—that is, the specific group value of a given conditional 'degree of belief'—is not a matter of mere convention? A group might agree on the basis of brainwashing, the reading of tea-leaves, or an infinite number of other means which plausibly have little, or nothing, to do with the truth. In short, internally consistent 'mob psychology' is still mob psychology!

Worse, a subjectivist could equally demand that (a specific class of) personal probabilities ought to be arrived at by 'critical thinking and thorough consideration'—that is, use this as a top-up rationality constraint, as discussed earlier. Nor can there be any necessary advantage in appealing to more than one person in making a decision without the specification of some further constraints. To put it bluntly, it is unclear what sort of advantage could be gained by asking two mass murderers to decide together whether killing is wrong, rather than asking one philosopher. Assuming moral realism, it is plausible that there is none whatsoever.

So while appeal to critical discussion and/or critical procedures looks fine on the surface, it only papers over some yawning cracks. Until we have an answer as to exactly what arriving at decisions by critical discussion can achieve, and precisely how this can result in better decisions than personal ones—with respect to what is actually the case—we have not solved the problem.

How to proceed? The first thing to note is that we are not going to refute scepticism via developing this interpretation of probability, at least not without making some robust metaphysical assumptions. (Who, indeed, would expect that we could?) The underlying problem is, after all, as old as philosophy itself. Why should what we think—as individuals or groups—have anything to do with some 'external world', even assuming that there is one (and there are 'deep truths' concerning it)?

Moreover, our radical fallibility is in no sense being questioned. A group can get things wrong, even repeatedly and systematically, in exactly the same way that an individual can. It's important to be clear about this, and trim our aspirations accordingly. But we shouldn't be too quick to draw the conclusion that the intersubjective and subjective views are on precisely the same footing, because our actual practice reflects the fact that we often don't take personal decisions to be as trustworthy as group ones. See Pettit (2006) on majority testimony, for example.

So why and when should we consider a group decision to be superior to an individual one? As a way into finding an answer, consider why it is that some journals employ two, or even three, referees. Is it just a matter of selecting an appropriate sample size to get a trustworthy measure of opinion from the set of philosophers (or philosophers of X, where X is the subject area of the piece)? This would make sense, but it does not strike me as the point; in fact, it would be a great shame if this were so. Instead, it seems that the purpose of the exercise is to determine the quality of the paper, and to isolate any errors therein. This is why a one-line report—'In my opinion, this is unsuitable for publication!'—does not provide sufficient grounds to reject a paper, whereas a careful and considered report is something for which any author can be grateful. Most of us really do revise our work with the truth in mind. We do not blithely agree with whatever a reviewer says, even if there is a serious promise of publication on the basis of appropriate revisions.

In support of the view that criticism via reviews improves quality, consider the case of *Atmospheric Chemistry and Physics* discussed by Koop and Pöschl (2006). It was ranked twelfth out of 169 journals in 'Meteorology and atmospheric sciences' and 'Environmental sciences', in terms of ISI impact factor, just three years after its foundation. Yet it rejected less than one in five submitted papers. What made this possible is an open review system. In the first stage, after only 'a rapid prescreening' (ibid.), submitted articles are published on the journal's website, and comments are invited both from appointed reviewers (who may choose to remain anonymous) and other interested scientists (who are asked to sign their contributions). The author is allowed to reply in public so that a dialogue can ensue. The second stage—revision and review in the traditional manner—only takes place two months later (at least). Koop and Pöschl conclude that: 'collaborative peer review facilitates and enhances quality assurance' (ibid.).

It seems correct, moreover, that a good editor doesn't merely compile the opinions of referees and adopt blanket policies such as 'One recommendation to publish and one recommendation to reject results in rejection'. On the contrary, the aim is to reach a considered consensus—the decision being informed, but not determined, by the content of the referees' reports—on the understanding that this achieves a significant epistemic task. It may increase the (objective) probability of ruling out papers containing dishonest claims, for instance. And editors do have a role to play in detecting misconduct, as Fox (1994) argues.

We can also see that bad decisions are liable to be made when members of the group are either incompetent with respect to the intended task, or are trying to 'play the group' for some personal advantage which precludes their (genuine) participation. In the first case, a review might be written by a new PhD student who is not well grounded in the relevant literature. In the second, a reviewer might have a personal grudge against an (easily recognisable) author.

But the competence of scientists is tested by a variety of standard procedures; successful contributions to science by way of publications and/or experimental activity, often resulting in an accredited professional qualification, are required for participation in serious research. So is testimony from other more established—or 'corroborated'—scientists, for example supervisors. The testing never stops; each and every putative contribution is double-checked. And what's more, honesty is also tested. To be caught falsifying one's results is fatal for one's scientific career, and the penalties are so great because accepting testimony, when it comes to research, is a norm. It's usually safe to take the author's word, given some admittedly contingent features of our academic community.

So if a more informed decision is liable to be a better decision, it might seem that an intersubjective probability reached by a group of competent individuals (e.g. 'scientific experts' in a given field or fields), sincerely working together (e.g. to find the truth), should be preferred to a subjective probability.[17] Yet there is an important objection to this line of argument. Why can't just one competent (and sincere) individual collect all the relevant information—for example, one 'lead scientist' simply interview the others, and/or consult all their papers—and thereby arrive at just as good a decision? (Let's forget practical constraints, in order to strengthen the objection.) In fact, this question raises a doubt as to where the subjective interpretation ends and the intersubjective one begins. We have two choices: to say that decisions informed by others are nevertheless properly subjective, or to admit them as intersubjective.

As counterintuitive as it may initially seem, I believe that the second option is the correct one, even when we take an individual being informed by the written research of others. Such work is, after all, the testimonial product of their inquiry, and to pool it is to take into account the products of our inquiry to date (or as Popper might put it, our 'objective knowledge').[18] As Lipton (1998, p. 1) puts it:

> At least most of the theories that a scientist accepts, she accepts because of what others say. The same goes for almost all the data, since she didn't perform those experiments herself. Even in those experiments she did perform, she relied on testimony hand over fist: just think of all those labels on the chemicals. Even her personal observations may have depended on testimony, if observation is theory-laden, since those theories with which it is laden were themselves accepted on testimony . . . We live in a sea of assertions and little if any of our knowledge would exist without it.

Indeed there is a burgeoning literature on testimony—see Pritchard (2004) and Lackey and Sosa (2006) for recent surveys—which recognises it as a significant epistemic phenomenon in its own right. Diller (2008, p. 421) has recently tackled testimony from a critical rationalist perspective, and shown how this is compatible with the rejection of evidentialism discussed in the previous chapter: 'None of the information acquired from testimony is justified in any way; it is accepted until there are reasons to reject it'.[19]

Even if I am wrong that the ubiquity of testimony supports an intersubjective understanding of many probabilities, moreover, it is plausibly the case that the greater the interactivity and participation when it comes to determining probability assignments, the better. First, two-way interaction (and three-way interaction, and so forth) can result in the production of ideas that would not otherwise have arisen.[20] For our purposes, this is particularly important when it results in a novel synthesis of, or new means by which to conceptualise/interpret, results—understanding new ways that particular theories can account for agreed phenomena, for instance.

Second, error correction is improved in larger groups. For instance, there are often circumstances in which individuals would agree provided that their mistakes were ironed out. In a simple case, a conference participant may have a mathematical error in their paper pointed out by an audience member. In a somewhat more complicated scenario, a scientist might be labouring under misconceptions about a theory in a different domain from the one in which he primarily works, in advancing a new hypothesis. And upon being informed that he is in error by an expert in this different field, he may simply accept the testimony. We may say that scientists (and experts in general) are often disposed to believe the same things—to use the terminology of Audi (1994)—but often need to have these dispositions activated; and a critical discussion on the relevant subject matter is an excellent way to do this. This is not to deny that error correction can be achieved by a lone researcher. It's just that lone efforts will have a greater chance of containing errors than group ones, *ceteris paribus*.

On a related note, consider the accuracy of well-accessed entries on Wikipedia, which are open to editing by any member of the public, as against those written by experts for *Encyclopaedia Britannica*. According to Giles (2005, p. 900): '[There are] numerous errors in both encyclopaedias, but among 42 entries tested, the difference in accuracy was not particularly great: the average science entry in *Wikipedia* contained around four inaccuracies; *Britannica*, about three'. And if we were to exclude Wikipedia articles which had been wilfully damaged—where the aim of one or more of contributors was not truth, that is to say—this figure would presumably alter in Wikipedia's favour. After Kuhn, it is hard to deny that the community nature of scientific endeavour is significant, and for more reasons than those enumerated here. But what I have tried to suggest is that this is for epistemic reasons—in addition to pragmatic concerns, for example, that 'the present "neofeudal" organization is actually a brake on efficiency' (Gillies 1991, p. 529)—at least when it is compared to an individualistic approach to inquiry. In fact, this is consistent with ideas that Popper didn't have much truck with; about the importance of exemplars (Kuhn 1977) in particular, which will be touched on in Chapter 6.

What's important to remember, in what follows, is that this discussion has nowhere supposed that corroboration is superior to Bayesian confirmation. Yet I shall now endeavour to show that this is a consequence.

4. INTERSUBJECTIVE CORROBORATION VS. INTERSUBJECTIVE BAYESIAN CONFIRMATION

Thus far, we have seen that 'intersubjective probabilities'—at least in the variant of the notion that I am interested in—should not be understood to include any group probabilities which are not the result of appropriate critical, and co-operative, activity. We have seen that the competence and character of the participants is significant, and that it is therefore important for there to be procedures in place by which reliably to select suitable individuals, and test their suitability to remain a part of the collective. But we are still faced with the question of which factors scientists should be considering; of what their shared values should be, and of whether these can be 'correct' in some sort of objective sense. This has been a concern of Kuhn (1977, ch. 13), van Fraassen (1980, pp. 87–89), Laudan (1984), and Worrall (1988).

In contrast to Popper's way of doing things, Kuhn attempts to explain the progress of science by specifying factors which we can apply in order to assess the relative status of competing theoretical frameworks: the theoretical virtues of 'accuracy, consistency, scope, simplicity and fruitfulness' (Kuhn 1977, p. 321). Kuhn also takes these factors to be significant with respect to both theory choice and theory construction. In his words (ibid., p. 335): 'If the list of relevant values is kept short . . . and if their specification is left vague, then such values as accuracy, scope and fruitfulness are permanent attributes of science'.

This approach is problematic because the predilections of individual scientists will have a serious effect on how they weight such values, even assuming that they can all agree on what they are. So even if it is right that 'everyone will readily agree that simplicity, informativeness, predictive power, [and] explanation are . . . virtues' (van Fraassen 1980, p. 8), this is simply not enough. An instrumentalist might inveigh against Bohm's interpretation of quantum mechanics, and focus on the supposed 'baggage' of the quantum potential in comparison to the Copenhagen alternative: in doing so, she would favour simplicity. The realist might rejoin by emphasising the informative nature of Bohm's interpretation; how it avoids the measurement problem and explains the classical limit (and so on; see Cushing 1994). So even putting to one side the fact that these virtues might be merely pragmatic, there is not liable to be much agreement, let alone principled agreement, on values for P(*h*,*b*)—the Bayesian 'prior probability' of a theory. This is a serious problem for *intersubjective* Bayesianism, even as a descriptive account of science. As Worrall (1988, p. 269) puts it:

> Views about aims and goals seem altogether more ephemeral, more 'philosophical' than judgements about which scientific theory is presently best supported empirically.

It is in virtue of this, we might think, that we ought to focus on the notion of empirical success *qua* resilience to testing. And sure enough, Gillies (1991, p. 530) suggests that there is greater ease with respect to reaching agreement on this:

> We want, as far as possible, to ensure that our confirmation function is based on intersubjective probabilities which are consensus probabilities of the whole relevant scientific thought collective. In this way we can achieve general agreement in judgement as to how the competing research programmes are progressing. But this means that we should try to confine ourselves to probabilities like P(*e*,*h*&*k*) and P(*e*,*k*), and try to avoid the prior probabilities P(*h*,*k*) of the Bayesians.[21]

Concerning the extent to which a given theory suggests a peculiar observation, we can reasonably expect a discussion between scientists to serve the function of error correction, discussed previously, rather than the production of an uneasy compromise about aesthetic differences. Compare and contrast this, again, with Kuhn's (1970b, pp. 237–238) emphasis on the significance of values, despite his agreement on the importance of group activity:

> [T]ake a group of the ablest available people with the most appropriate motivation; train them in some science in the specialties relevant to the choice at hand; imbue them with the value system, the ideology, current in their discipline; and, finally, let them make the choice.

Naturally, it is possible to rank the importance of virtues by stipulation: to say that simplicity outweighs scope, that accuracy outweighs simplicity, and so forth. However, it is not plausible that there is a single, universally correct way in which to do this in order to achieve the aim of science (whether it be truth, empirical adequacy, or even instrumental success).[22] At best, it would appear that which values *should* take precedence, with reference to the aim of science, is a matter of context. Disagreement on priors therefore seems natural, and will often be difficult to resolve.

It must be emphasised that our apparent ability to reach considered consensus on evaluations of P(*e*,*hb*) and P(*e*,*b*), as against those of P(*h*,*b*), might nevertheless fail to be of any deep epistemological significance. Perhaps our liability to agree is just a matter of psychological fact? This can be accepted, because what is at stake is the relative prospects of competing approaches—intersubjective Bayesianism versus intersubjective corroboration—for (the explication of) the rationality of science. If you think that our judgements about whether a theory is suggestive of a particular observation are not generally reliable—and even that we need not behave as if they are, to enable scientific inquiry—both approaches will be unacceptable. But this will not expunge the distinction between the two types of judgement involved. Evaluations of P(*h*,*b*) are clearly axiological. Evaluations of P(*e*,*hb*) and P(*e*,*b*) are not.

ACKNOWLEDGEMENTS

This chapter is based on Rowbottom (2008c).

4 Corroboration, Tests, and Predictivism

In the previous two chapters we have seen how a corroboration-based account of scientific method can stand without appeal to induction, and that corroboration values should be based on intersubjective probabilities formed by suitable groups and procedures. The discussion has emphasised the significance of testing both ideas and people, and has suggested that these activities are at the heart of the scientific enterprise. Just as a theory will not make it into a textbook (except perhaps as an example of a mistake) without having stood up under fire, so a scientist will not make it into the laboratory without having jumped through several professional hoops. Of course, accidents do happen. We have some bad theories, and some bad scientists. But we do our best as a collective—and with a margin of success, it would seem—to minimize these. (Although it should be noted that it can sometimes serve collective ends best to admit people who violate the canons of individual rationality; I discuss this idea in Chapter 6.)

There are still several outstanding problems for a corroboration-based account of scientific method, however. One especially noteworthy example, which will be discussed in depth in the next chapter, is Duhem's thesis that a hypothesis cannot be tested in isolation. In the present chapter, we will discuss three further problems. First, when exactly should an observation statement be understood to be worthy of counting as an *e* in the corroboration function? We have already seen that Popper thought that not just any old observation statement will do, but instead that we should only use the results of tests (at least to positively corroborate theories), and then only the severest tests at that.[1] As we will see, however, this view is not so easy to sustain as it may first appear. One key problem we will encounter is what I have elsewhere called 'The Problem of the Big Test': that the severest test of any hypothesis is to perform all possible tests of said hypothesis (when 'possible' is suitably interpreted).

Second, and on a related note, should we prefer theories which make novel predictions to those that only accommodate facts already known? One way of seeing the significance of this, with regard to the corroboration function, is to consider that we could calculate corroboration values on the basis of 'as-if' background knowledge, or background knowledge past. Imagine *e* was discovered before *h* was formulated. We could still calculate a value such as P(*e*,*hb*) – P(*e*,*b*) for the state of scientific knowledge, *b*, previous to the discovery of *e* (and so imagine that *h* had been formulated earlier than it actually was). Is it appropriate to do this?

The third and final question we will consider is whether sincerity matters in the testing process. Musgrave (1974b) has challenged this view, arguing that an experiment can serve as a test of a theory whether or not the person performing it intends to refute said theory (and even that casually discovered information can positively corroborate a hypothesis), although Popper appears to have thought otherwise. Who is right?

1. A DISCONTINUOUS VIEW OF CORROBORATION

According to the standard view of Popper's account of corroboration, which I shall call the continuous view, a hypothesis has a degree of corroboration at any point in time subsequent to its first test. As relevant new evidence comes in, that is to say, the degree to which a hypothesis is corroborated may (and often will) change; but it will never cease to have a corroboration value. The only additional proviso, recall, is that the tests must be intentional: '*C*(*h*,*e*) must not be interpreted as the degree of corroboration of *h* by *e*, unless *e* reports the results of *our sincere efforts to overthrow h*' (Popper 1959, p. 418). This is the same as saying that information derived from sources other than tests, e.g. accidentally acquired observation reports, cannot in any way support—although, Popper would presumably maintain, can still refute—a hypothesis.[2] Testing a theory involves more than merely 'applying it' or 'trying it out'; instead we must examine 'cases for which it yields results different from those we should have expected without that theory, or in the light of other theories' (Popper 2002 [1963], p. 150).

So according to Agassi (1959, p. 317), for example, 'Corroboration of a theory is merely an appraisal of the way it stood up *so far* to severe criticism'. And according to O'Hear (1975, p. 274): 'degree of corroboration is . . . an account of a theory's performance up to a certain time and compared to other available theories; it is therefore contingent on the tests we have been able to devise and execute for it and the other theories suggested by our ingenuity'. Corroboration has also been understood similarly elsewhere, by friend and foe of Popper's philosophy alike, e.g. by Cohen (1966) and Gillies (1971). A rather different understanding, however, might be suggested on the basis of the following passage:

> I must insist that *C*(*h*,*e*) can be interpreted as degree of corroboration only if *e* is *a report on the severest tests we have been able to design.* (Popper 1959, p. 418)

Although these words have been noted and quoted before—in fact, Keene (1961, p. 86) summarises them as follows: 'In other words, we cannot finally arrive at our assessment of the degree of corroboration until the results of severe testing are at hand'—their full significance appears to have been missed.[3] The crucial point is that Popper uses 'design', rather than O'Hear's 'devise and execute'. It follows that we might find ourselves in situations where a hypothesis fails to have a degree of corroboration because the severest tests we have been able to design have not yet been performed. (That is, on the reasonable assumption that we can only have a report on the outcome of a test after it has actually been performed.) But this means that a hypothesis which has a corroboration value today need not have one tomorrow, and that this may occur not because of new empirical evidence, but *simply because someone has devised a new test for the hypothesis which is more severe than any that has gone before.* Corroboration would be discontinuous in the following sense. If we were to plot corroboration values against time, we would (sometimes) generate graphs with essential discontinuities (rather than mere steps).

This reading—which I will henceforth call the discontinuous view—does appear to be contradicted by some of Popper's comments about corroboration elsewhere, e.g. concerning '*the severity of the various tests* to which the hypothesis in question can be, and has been, subjected' (Popper 1959, p. 267) and 'values which are logically derivable from the theory and the various sets of basic statements accepted at various times' (ibid., p. 275), and there is no mention of anything similar in his later discussion of corroboration (Popper 1983, ch. 4).[4] Nevertheless, it's an interesting proposal in its own right. What's more, as we shall see, criticising it leads to a result that has wider implications.

Before launching into this criticism, however, we should address one final claim of Popper's concerning degree of corroboration of a given hypothesis, namely that it cannot move from a negative value to a positive one:

> A corroborative appraisal made at a later date—that is, an appraisal made after new basic statements have been added to those already accepted—can replace a positive degree of corroboration by a negative one, but not *vice versa*. (1959, p. 268)

This is inaccurate, because an appraisal made at a later date need not be one where new basic statements have simply been added to those already accepted.[5] In fact, it is possible to remove basic statements too, e.g. on discovering that a scientist's report of an experiment was fraudulent or that a piece of apparatus was malfunctioning during an experiment. So a severe test which a hypothesis appears to have failed may turn out to have been inconclusive. At that point we must presumably revert to the previous corroboration value for the hypothesis according to the standard view of corroboration, although we would have to suspend judgement (i.e. the hypothesis would cease to have a corroboration value) according to the discontinuous view outlined earlier.

Moreover, we can imagine having new basic statements which cast doubt on previous experimental results *without* resulting in the rejection of any previous basic statement(s) in particular. Imagine we discover that a piece of electronic measurement apparatus has an intermittent fault, for example, so that only some of our previous tests of a hypothesis have been genuine. What we should do according to the standard view—what, that is to say, the value of corroboration then becomes—is unclear. The discontinuous view, however, results in a clear recommendation. As the severest tests of *h* that have been designed have not been (properly) performed, there is no corroboration value for *h*. This may be an advantage for the discontinuous view.

2. THE PROBLEM OF THE BIG TEST

The most obvious objection to the discontinuous view of corroboration is that our ability to design tests far outstrips our ability to perform them. It may be possible to devise a test using an imagined particle accelerator which could only be built far in the future, if at all, for instance. And even if a time did come when it could be built, it seems reasonable to suppose that someone would by then have devised an even more severe but completely impracticable test, and so forth. It therefore looks as if there would be no corroboration values, ever, for many—if not all—hypotheses; that is, if the discontinuous view were correct. To recapitulate, this is because its core requirement is that *e* is '*a report on the severest tests we have been able to design*', rather than a report only on the severest tests that we have been able to design *and* perform. (And naturally, we cannot—provided time travel is impossible—have a report on a test that has not yet been performed.)

One way to address this concern is to suggest that 'design' has a somewhat different connotation from 'devise'. And just as to devise a rocket is not necessarily to design a rocket, so it might be said that to devise a test is distinct from designing a test. To return to the previous example of the bright spot, for instance, we might say that Poisson merely devised a (kind of) test which he strongly anticipated that the wave theory of light would fail. We might say that Arago designed the actual severe test, i.e. peculiar experiment, shortly before performing it.

Even accepting this, however, the present problem is far from solved in all cases. Consider the hypothesis that the coin in my pocket is fair, i.e. gives a heads result as often as a tails result when flipped by me.[6] The severest test I can design involves flipping the coin as much as I possibly can, while allowing as little time as possible for activities required to keep the experiment running—eating, drinking, taking light exercise, and so forth! Indeed this test is certainly *possible* in a sense that other conceivable (or merely 'devisable') experiments, involving infinite numbers of flips, are not. I could leave my wife, abandon my daughter, and give up my job. I could devote my life to flipping that infernal coin. The hypothesis would have a corroboration value only upon my demise, at best, according to the discontinuous view. This is because this severest test would only be completed upon my death, and therefore the relevant evidence from that test would only be available—to allow a judgement on the value of $P(e,hb) - P(e,b)$—at that point.

This rather bizarre scenario suggests a more general objection, namely 'The Problem of the Big Test'. In order to set it up, consider first that although 'there is something like a law of diminishing returns from repeated tests (as opposed to tests, which, in the light of our background knowledge, are of a *new kind*, and which therefore may still be felt to be significant)' (Popper 2002 [1963], p. 325), it is nevertheless true that the more we repeat a test procedure, the more severely we test the relevant hypothesis.[7] As Popper (1983, p. 248) puts it, 'a multiple test is more improbable—and accordingly also *more severe*—than its component tests'. Nothing like enumerative induction is required here. In the coin-flipping example, as in almost any other conceivable testing procedure for a suitably universal hypothesis, the background conditions will differ each time the coin is flipped. And we cannot be sure that some of those changes will not have a profound effect on the results.

Second, we need only recognise that what counts as a series of tests rather than a single test appears to be a matter of convention. Imagine that I claim yesterday to have performed two tests of the hypothesis that whenever I flip a peculiar coin, it always lands on heads. Furthermore, I report that the hypothesis was only falsified by the second test, which was considerably more severe than the first. Now what actually happened, let's say, is that in the morning I flipped the coin once (and it landed on heads), and in the afternoon I flipped the coin nine times (and it landed on tails the third time). Would it be misleading for me to say that I performed just *one* test of the hypothesis on that day, which involved flipping the coin ten times (or even just four times)? The intuitive answer lies in the negative. One could always suggest that each flip should be understood as a separate test. But this is a thoroughly counterintuitive way to understand such scenarios, as becomes clear when we consider instead the hypothesis that a given die is fair. There is no minimum number of rolls that clearly counts as a test in its own right, although it is evident that a single roll does not count as a test at all and that the more rolls we perform, the more severely we are testing the hypothesis. It would certainly be perverse, for instance, to insist that to roll the die three hundred times on one day is to perform thirty tests of equal severity, rather than just one test! And even if a principled way can be found of distinguishing between tests in some circumstances, it nonetheless seems that many different tests are often considered to amount to a single test because they are directed to the same end. The driving test in the United Kingdom has both a practical and a written component, and one can fail because of one's performance on either. We nevertheless consider it to be a single test of one's driving ability.

We now arrive at the Problem of the Big Test. If 'a multiple test is . . . *more severe* . . . than its component tests' (Popper 1983, p. 248) and a multiple test may legitimately be considered to be a single test in its own right, then there is just one severest test for any given hypothesis at any given time, which involves performing all possible (component) tests that can be designed. Since this will generally be impossible, hypotheses will never attain corroboration values according to the discontinuous view.

It might be objected, perhaps with reference to some of the pragmatic aspects of Popper's thought, that we nevertheless have a pretty good intuitive idea of what counts as a single test, and that this suffices for all practical purposes. To return to the previous example, for instance, it might be thought to be clear that a driving test is one kind of test of one's driving ability, whereas one's performance on public roads (subsequent to passing that test) is another. Still further, it might be said that driving in a city environment (with lots of traffic and wide lanes) is a different test from driving in the countryside (with less traffic but narrow lanes). My response is twofold.

First, even if we accept that our intuitions are *sometimes* sufficient, as perhaps they are in these cases, it does not follow that they always are. Think now of whether driving downhill is a different test to driving uphill (on the same short journey), or whether driving with the radio on is a different kind of test to driving with the radio off (and whether driving while listening to a CD, rather than a radio station, makes any difference). Think, further, of whether we should take each second's worth of driving to constitute a distinct test, rather than each minute's worth, or each year's worth. (Remember also the previous die rolling example.) Here, I contend, our intuitions are either silent or divergent.

Second, even if the previous objection fails, note that we are supposed to rank these tests with respect to severity (if we just want to perform the severest test). Consider again the test of driving in the city (for one hour) as against the test of driving in the country (for one hour). Which is the more severe? I contend that this is not clear. But what does seem to be clear is that a driver who has undertaken both tests will usually have been more severely tested than a driver who has only undertaken one. And surely no one would object to us considering 'driving in the city (for one hour) and driving in the country (for one hour)' to be a test of one's driving ability? There is no obvious argument that a test cannot be composed of tests, and ultimately this is all that is required for the Problem of the Big Test to retain its force.

3. A LESSON FOR BOTH ACCOUNTS OF CORROBORATION

Prima facie, the Big Test objection appears simply to rule out the discontinuous account of corroboration. But there is also a lesson to be derived from it for the standard account, namely that the severest test of any given hypothesis which has been designed and performed can be understood simply to be the sum of *all* tests which have been designed and performed. In effect, this means that to say '$C(h,e)$ can be interpreted as degree of corroboration only if e is *a report on the severest tests* [that we have devised and performed]' (Popper 1959, p. 418) amounts to saying that '$C(h,e)$ can be interpreted as a degree of corroboration only if e is *a report on all the tests of h that we have devised and performed*'.

Note that when we consider the crucial measure of corroboration, namely $P(e,hb) - P(e,b)$, it becomes clear that this is not simply to demand, with Carnap (1962, p. 211), that 'the total evidence available must be taken as a basis'. In so far as all the available information at any point in time should be in either e or b, this is evidently true.[8] But it is compatible with letting the reports of lax tests fall into b rather than e, which is precisely what we should not do according to my findings. As such, the complete recommendation (for the standard continuous view of corroboration) is '$C(h,e,b)$ can be interpreted as a degree of corroboration only if e is *a report on all the tests of h that we have devised and performed* and b contains all the other information we have.'

This is may lead us to wonder why Popper wrote of 'severest tests' at all. Did he have something like the discontinuous view in mind, after all? If so, then the introduction of practical considerations, relating to utility in particular, would appear to be unavoidable in order to avoid the problem of the Big Test. To be specific, we would have to consider only a subset of those tests we have been able to design; those that could be performed in the near future without prohibitive expense and/or difficulty, for instance. This, in turn, would mean that convention would ultimately have to determine whether a given hypothesis was taken to have a corroboration value or not. There is no obvious reason for which Popper would have objected to this proposal; conventions can be principled.

4. ACCOMMODATION VS. PREDICTION

A further issue, which remains of interest in the current literature (e.g. Barnes 2005; Hudson 2007; Harker 2008), is raised by the Problem of the Big Test. If we must (or even if it is simply legitimate to) consider all tests, past and present, as one Big Test, then might we not understand such testing in terms of accommodation, rather than prediction? It is not hard to see how this idea arises, because we would be treating old data, sometimes very old data, as if it were new. This may seem to be at odds with Popper's (2002 [1963], pp. 47–48) dictum that: 'Confirmations should count only if they are the result of *risky predictions*; that is, to say, if, unenlightened by the theory in question, we should have expected an event which was incompatible with the theory—an event which would have refuted the theory'.

Before considering this issue in further depth, it will be helpful to distinguish clearly between two forms of possible accommodation, as outlined, for example, by Scerri and Worrall (2001, p. 424). According to the first, which I shall henceforth refer to as 'weak accommodation', a theory accommodates some fact if the fact was known before the theory was generated but not used in the construction of the theory. According to the second, which I dub 'strong accommodation', a theory accommodates some fact only if the fact was known before the theory was generated *and* used in constructing the theory.

Now as hinted in the previous section, during the brief discussion of total evidence, this is a situation where it is illuminating to consider the core of the formal measure of corroboration, $P(e,hb) - P(e,b)$. (Recall that the denominator of Popper's corroboration function is only there for the purposes of normalisation; as such, we can put this to one side in the interest of clarity.) The issue is which propositions we should put into *e* and which propositions we should put into *b*. The suggestion might be that all test data should fall into *e*, and that *b* should represent either our current background knowledge minus any data from previous tests, or our previous background knowledge at the time that *h* was initially proposed.

A concrete example will help to explain. Let's imagine that we want to determine how well corroborated the theory of special relativity is, and are considering all the tests of the theory to constitute a single Big Test. If we follow the first path, we may imagine a possible world in which we don't know the results of any of the tests of special relativity which have actually been performed (e.g. due to Hafele and Keating 1972 and Allan et al. 1985) since the theory was proposed—if desired, we can imagine a scenario in which these tests are yet to be performed—and in which no additional (i.e. non-actual) tests of special relativity have been performed, but in which we have all of our other present knowledge. We may then imagine doing the tests that have actually been performed, and getting the same results as we actually have. That is to say, *b* will be based on our current background knowledge: call this $b^{\dagger}$. If we follow the second path, however, we may imagine we were around at the point at which the theory of special relativity was proposed (and only in possession of the scientific knowledge of that time).[9] Again, we can then imagine performing all the tests that have actually been performed, and getting the same results as we actually have. In short, *b* will be based on our past scientific knowledge (at the time of the generation of *h*): call this b^{*}.[10]

Irrespective of whether we prefer to use $b^{\dagger}$ or b^{*}, there is no need to accept strong accommodation as a form of corroboration. In both cases, nothing counts as a test unless it was performed *after* the generation of the relevant hypothesis (*h*), which happens to be special relativity in this instance. (And it is a requirement that *e* must be a report on a test, or tests.) This leaves us only with the worry about weak accommodation.

Now with a little further thought, it emerges that the second of the aforementioned strategies, involving b^*, is unacceptable. The main reason is that some of the content of b^* is widely accepted, here and now, to be false. So it appears curious to suggest that we should be interested in the value of $P(e,hb^*) - P(e,b^*)$, in the context of assessing the present worthiness of *h*, when this is so. The other reason is that much of the knowledge required to perform the tests which have now been performed was simply not available when special relativity was first proposed. The test of Allan et al. (1985), for example, uses the Global Positioning System. But in 1905, we were not even in the position to put satellites into orbit!

We are therefore left with the first strategy, involving the use of $b^\dagger$. Need we accept the value of weak accommodation? I believe that the answer lies in the negative if we require that '[e]very genuine *test* of a theory is an attempt to falsify it' (Popper 2002 [1963], p. 48) and recall that '$C(h,e)$ must not be interpreted as the degree of corroboration of *h* by *e*, unless *e* reports the results of *our sincere efforts to overthrow h*' (Popper 1959, p. 418). In short, *e* can only be the report of the results of a genuine test of *h* if *e* is (temporally) novel relative to *h*; that is to say, if *h* was proposed before *e* was known (or believed to be true).

However, this does not erase the significant difference between using $b^\dagger$ (which is suggested by the Big Test) and other possible strategies for calculating C, e.g. putting the results of each test of *h* prior to any current test into *b* (so that all old *e* is always part of current *b*) and just selecting one historical calculation of C, e.g. according to severity criteria, to provide the definitive corroboration value for *h*.

5. IS PREDICTION MORE IMPORTANT THAN EXPLANATION?

One might nevertheless question whether it is correct to say that '[c]onfirmations [or rather corroborations] should count only if they are the result of *risky predictions*' (Popper 2002 [1963], p. 47). One kind of argument in favour of this view, which Snyder (1994) associates with Whewell (1860), is based on inference to the best explanation. The idea is that the best explanation for a successful risky prediction by a theory is the truth of the theory, rather than luck (or chance). However, a critical rationalist would not want to appeal to such an inferential strategy. As we saw in Chapter 2, she would maintain, rather, that corroboration is simply *not* an indication of the truth, truth-likeness, or even empirical adequacy of a theory.

Snyder (1994, p. 463) therefore suggests that Popper might employ the following alternative argument:

> The best that any evidence can do . . . is to corroborate a theory. On this view, only predictions can count as evidence, because only predictions

> are "potential falsifiers" of a theory. If a theory explains a known fact, this information cannot potentially falsify the theory because it is already known that the fact explained is true. But if a theory predicts an unknown fact . . . then it can function as a potential falsifier: we can attempt to observe whether this phenomenon does, in fact, occur. . . .

It seems to me, however, that this is incomplete (at best). Even if it is true that only predictions are 'potential falsifiers', the real reason that evidence noted in the past is not so significant, for the critical rationalist, is that what we have recorded (and/or noticed) may be inappropriately selective. That is to say, situations where the theory has been falsified may have been overlooked.

But what is the argument against predictivism? Brush (1989) discusses the example of the precession of the perihelion of Mercury, which could be accounted for by relativistic considerations (as Einstein showed) but not via Newtonian mechanics. He suggests:

> [A] successful explanation of a fact that other theories have already failed to explain satisfactorily (for example, the Mercury perihelion) is more convincing than the prediction of a new fact, at least until competing theories have had a chance (and failed) to explain it. (1989, p. 1127)

Despite the descriptive language in this quotation, let us assume that this how things *should* be on Brush's account (as Snyder 1994 does). The appropriate response is twofold. First, relativity would hardly have been acceptable at all if it weren't for the fact that it made *some* new predictions; and as such, explaining the previously unexplained is not *sufficient* for a novel theory to be convincing (even when no rivals explain anything else that it cannot explain). Second, and crucially, a theory's ability to explain something previously unexplained may be a reason to think it worthy of further investigation (i.e. to employ it in some theoretical contexts) when it also makes novel predictions. But this is not a reason to think it corroborated or confirmed. So the worry is that Brush conflates corroboration (or worse, confirmation) with preference from a peculiar theoretical perspective. In short, what counts as a reason for taking T* as a serious *potential* successor to T (and therefore worthy of more detailed empirical investigation) is not the same as what counts as a reason for taking T* to *be* a successor to T. It may help to recall some of the discussion from Chapter 2 here. Falsified theories can be useful in some theoretical and practical contexts, especially when they nevertheless appear to be empirically adequate in a range of circumstances.

In short, we are impressed when a theory passes a test involving a risky prediction because it has resisted an attempt at falsification which we expected to be successful.

6. DOES SINCERITY MATTER?

One final criticism of Popper's view on corroboration is that the demand that '*e* reports the results of *our sincere efforts to overthrow h*' (Popper 1959, p. 418) in order for *e* to be applicable in the corroboration function is incorrect. Musgrave (1974b, pp. 577–578) advances this objection as follows:

> [T]he *sincerity* with which a test is devised and performed seems to be distinctively psychological, to depend upon the state of mind of him who performs it. Evidence which corroborates a hypothesis must not be taken into account, says Popper, unless the tester was sincerely trying to refute the hypothesis in question . . . But how are we to find out that this requirement *is* met? A report of an experimental test would, it seems, have to be accompanied by a psychological report that the tester was sincere before it could be taken seriously as an argument! . . . It is perfectly possible for a severe test to be performed by one who does not sincerely want to refute the theory in question. The test may hope, and try, to confirm the theory—he may try to do so in a spectacular way, by showing that the theory successfully predicts a new effect. His test will be a severe one (whatever its outcome). It is probable that sincere critics will be more *likely* to produce severe tests, but sincerity is neither necessary for severity nor sufficient.

Musgrave has a point. In fact, it is arguably possible for someone to perform a test of a hypothesis without even realizing that they are so doing. Imagine that my personal probability for $P(e,hb)$ is equal to my personal probability for $P(e,b)$ and I perform an experiment (which I did not intend to test h) where I find e. $P(e,hb) - P(e,b)$, for me personally, is zero. But clearly this is consistent with $P(e,hb) - P(e,b)$ being close to unity when we consider the intersubjective probability of the scientific community (or even the logical interpretation of probability rejected in Chapter 3). In fact, this argument may be strengthened by the recognition that I might consider only my own personal background knowledge of science (b_p), whereas I ought to be considering the community's (b); and clearly the value of $C(h,e,b_p)$ may be considerably different from the value of $C(h,e,b)$ even on a logical interpretation of probability. So should the community, on learning that I had no intention of testing h, decide that I have not succeeded in (unwittingly) corroborating it? *Prima facie*, this appears wrong.

In his response to Musgrave, Popper says that his references to 'sincerity' were supposed to highlight a significant methodological rule. Popper points out that the principle of total evidence discussed earlier—that is, 'the requirement that in assessing the degree of corroboration of a hypothesis, we consider *all* the available evidence' (Musgrave 1974b, p. 578)—is insufficient to ensure that the corroboration function is not misused. In his own words:

> Carnap thought that "total evidence" would give his formula some much needed security from misuse; that it would rule out situations where favourable or nonfavourable evidence had been carefully selected in order to obtain just the desired value for his degree of confirmation. I hoped to point out, with my reference to sincerity, that this is hardly an adequate safeguard against such "fixing"; that we need only shut our eyes at appropriate moments (when we fear the evidence may be unhelpful) to make this "total evidence" arbitrary and biased . . . "sincerity" therefore was obviously not meant in a psychologistic sense. (Popper 1974b, p. 1080)

The way that Popper (1983, p. 236) explains his position later is therefore, perhaps, more helpful: 'Only if the most conscientious search for counterinstances does not succeed may we speak of a corroboration of the theory'. Not only should we not rest content with old evidence, as we saw in the previous section, but we should also not rest content with what new evidence (e.g. reproducible effects) we find by accident. So perhaps we may moderate Popper's position by saying that my finding of *e* should certainly be of interest to the scientific community from the point of view of corroboration, although it would be legitimate to worry that the way I had come upon *e* might be such as to (potentially) preclude falsifying evidence for *h*. (The point is not just to test the veracity of *e*, but also the completeness of *e* in experimental context.[11]) So perhaps Popper would insist that the scientific community should repeat the experiment, and maybe even other related tests, before 'speak[ing] of a corroboration'.

ACKNOWLEDGEMENTS

This chapter is based on Rowbottom (2008b).

5 Corroboration and Duhem's Thesis

The question of how to deal with auxiliary hypotheses in confirmation theory is still live, and has recently been discussed by Strevens (2001, 2005) and Fitelson and Waterman (2005, 2007).[1] In essence, the core problem is as follows. How do we measure the (dis)confirmation of a theory, as distinct from the (dis)confirmation of the theory plus the auxiliary hypotheses (and/or other data) used to render it predictive?

However, little attention has been devoted to how this issue plays out when we consider corroboration as defined by Popper, i.e. focus on the measure $P(e,hb) - P(e,b)$, rather than confirmation of a Bayesian variety. This is important because Duhem's thesis is often said, without careful investigation, to provide the basis of an unassailable refutation of Popper's proposed methodology. As Nola (2005) puts it:

> A commonly cited obstacle to Popperian falsification is . . . the Quine-Duhem thesis in one or other of its several forms. This is something which . . . Popper recognised in *The Logic of Scientific Discovery* (section 18) when he confessed that we falsify a whole system and that no single statement is upset by the falsification. Popper seems to pay little further attention to the problem of how falsification, or even corroboration for that matter, may arise by piercing through any surrounding accompanying statements, to target a hypothesis under test.

In what follows, I will argue that Duhem's thesis does not decisively refute a corroboration-based account of scientific methodology, but instead that auxiliary hypotheses are themselves subject to measurements of corroboration which can be used to inform practice. I will also argue that a corroboration-based account is equal to the popular Bayesian alternative, which has received much more recent attention, in this respect.

1. FALSIFICATION AND DUHEM'S THESIS

Several well-known criticisms of falsificationism and/or critical rationalism hinge on Duhem's (1954 [1906], p. 183) thesis that 'an experiment . . . can never condemn an isolated hypothesis but only a whole theoretical group . . . [so] a "crucial experiment" is impossible'.[2] With respect to demarcation, for example, it may be said that no (or few) theories are falsifiable by, because they cannot be *directly* confronted with, experience (or more properly, observation statements).[3] From a methodological point of view, moreover, it would appear that we are in rather a quandary when we have evidence that is incompatible with a consequence deduced from a theory and auxiliary assumptions. When, precisely, should we take the theory to be falsified? As Strevens (2001, p. 516) says, the worry is that:

> when a conjunction of a central hypothesis and one or more auxiliary hypotheses is refuted, there is no principled way to distribute blame among the conjuncts, and thus that it is impossible to say to what degree the refutation disconfirms the central hypothesis.

Popper (1959, p. 83) does state that we should not use *ad hoc* hypotheses to defend theories from falsification: 'As regards *auxiliary hypotheses* . . . only those are acceptable whose introduction does not diminish the degree of falsifiability or testability of the system in question'.[4] But Popper's emphasis is rather curious, because presumably we should not use *ad hoc* hypotheses to falsify theories either. In fact, it appears that we simply shouldn't use *ad hoc* hypotheses in order to decide the fate of theories.

The question then arises as to what counts as an *ad hoc* manoeuvre. Popper (1983, p. 232) suggests that it involves a hypothesis 'which goes as little as possible beyond the facts it is expected to explain'. Yet it must be recognised that so described, such a hypothesis may sometimes be reasonable to adopt (or, at the very least, true). This is most obvious when the hypothesis is a singular statement—see the previous note—which might simply be the negation of a statement used in testing the theory. Consider, for instance, the hypothesis that a sample was contaminated because some of the equipment used to analyse it was not properly sterilised. Oftentimes, such hypotheses are not testable at all; it may be normal to sterilise equipment each time after it is used, but one might sometimes forget to do so.

There are also more complex theoretical examples of successful *ad hoc* hypotheses, such as Pauli's posit of the neutrino in order to explain the continuous energy spectrum of beta decay. The problem, in short, was that the amount of energy possessed by electrons emitted in beta decay was less than expected, given the changes in the nucleus, and not fixed. One option was, therefore, to challenge the notion that energy is always conserved. But Pauli instead suggested that beta decay involves another particle that can take some of the energy which would otherwise be possessed solely by the emitted electron. Leplin (1975) explains in depth that this posit was clearly *ad hoc*, when first introduced, for several reasons. (Popper [1974e] agreed, in a place that Leplin missed. See the next note.) Given the present focus on testability, the following is especially pertinent:

> Since the neutrino had not been detected, Pauli . . . assumed that it is extremely penetrating. Having no mass it would propagate with the velocity of light like the photon, but would have to be much more penetrating than photons of the same energy. There appeared little prospect that a particle satisfying these assumptions would be detectable. (Leplin 1975, p. 339)

Admittedly, what precisely counts as an *ad hoc* hypothesis is controversial; see also Grünbaum (1976).[5] But even if it were a simple matter to identify and rule out *ad hoc* hypotheses, cases would nevertheless occur where it was unclear whether the theory or the auxiliaries (or both) should be considered falsified by some observation statement. This leaves a problem that I will endeavour to solve in the remainder of this chapter.

I will do so by extending Popper's notion of corroboration to cover auxiliaries. In particular, I will endeavour to show that the following two theses, which are based on Bayesian equivalents advanced by Strevens (2005), can be motivated:

> 1*. When some evidence falsifies a conjunction of hypothesis and auxiliary, the 'blame' should be distributed between them roughly in proportion to their relative degrees of prior (independent) corroboration.
>
> 2*. When some evidence falsifies (or corroborates) a conjunction of hypothesis and auxiliary, the magnitude of the impact of the evidence on the hypothesis is greater the more (independently) corroborated the auxiliary.

2. CORROBORATION AGAIN

Since we have already covered the corroboration function in the previous chapters, I will not explain it again here! The key thing to note about the function, in what follows, is that it is fundamentally dependent on *b* in so far as all the probabilities in terms of which it is defined are conditional on *b*. That is to say, no probability relevant to determining C(*h*,*e*,*b*) can be determined without reference to *b*. This is a feature of all confirmation functions, including the 'one true measure' discussed by Milne (1996) and its competitor discussed by Huber (2008); so this should not be taken to be a peculiarity, let alone a weakness, of the corroboration function.[6]

Consider now Popper's (1983, p. 233) claim that:

> It amounts to adopting the uncritical attitude if one considers an event, or an observation (*e* say) as supporting or confirming a theory or a hypothesis (*h* say) whenever *e* 'agrees' with *h*, or is an instance of *h*.

We may now restate the criticism in the previous section as follows. Given Duhem's thesis, one might equally say: "It amounts to adopting an uncritical attitude if one considers an event, or an observation (*e* say) as undermining or refuting a theory or a hypothesis (*h* say) on the basis of some background information (*b* say) whenever *e* 'disagrees' with *h* in the presence of *b*, *even if* e *is the report of a sincere effort to overthrow* h!" Even if we accept that *e* is beyond doubt and/or is not susceptible to any serious criticism (at the point in time), the problem might lie with *b* rather than with *h*. And there would be nothing 'critical' about blithely ignoring this possibility!

Henceforth, let *A* denote the auxiliary hypotheses used in a test; *A* includes all those statements used to predict *e* in combination with *h*. And let *a* denote any individual auxiliary hypothesis; that is, any member of *A*. Popper (1983, p. 246) suggests that *A* should be 'sunk' into *b* in calculating corroboration values: 'By our background knowledge *b* we mean any knowledge (relevant to the situation) which we accept—perhaps only tentatively—while we are testing *h*.' This will include not only statements of initial conditions, but also 'theories not under test' (Popper 1983, p. 252). The content of *b* will not (usually) be exhausted by *A*.[7] What this means, in short, is that corroboration values are fundamentally dependent on auxiliary hypotheses. If we move from A_1 to A_2 then we will equally move from b_1 to b_2. And we will then be interested in $C(h,e,b_2)$ rather than $C(h,e,b_1)$. So, formally, to restate our problem in a third and final guise, we must recognise that even if $C(h,e,b)$ is minus one and we are absolutely satisfied that *e* is a true report of a test of (or genuine attempt to refute) *h*, it does not follow that *h* is false. The problem may lie in *b*, and more particularly in *A*.

Note that the same goes if we choose (by principled methodological convention) to say that any particular negative value of C (e.g. –0.75) counts as the falsification threshold. The need for a falsification threshold—a point at which the hypotheses under consideration are classified as false although they are not strictly incompatible with the observations made—is suggested by statistical hypotheses. For example, let our hypothesis be that the probability of a heads result on the flip of a particular coin (about which we have no prior data) is 0.5. We may stipulate that twenty consecutive tails results on the next twenty flips, *or any other similarly improbable result on the assumption that the probability of each flip resulting in heads were really 0.5*, would be sufficient for this hypothesis to pass the falsification threshold.[8]

3. A SOLUTION?

Our key question, now stated formally, is "What should we do when $C(h,e,b)$ has a value beneath the falsification threshold?" Is this a question we can answer, above and beyond saying that either *h* or *b* (or both) must go (without appealing to Duhem's notoriously vague 'bon sens')? More particularly, can we give an answer that is consistent with the spirit of Popper's philosophy?

At a rather obscure juncture, noted by Glymour (1980), Popper (1960, p. 132, fn. 2) claims that Duhem 'only shows that crucial experiments can never *prove* or establish a theory' but not that 'crucial experiments cannot *refute* a theory', and goes on to suggest:

> Duhem is right when he says that we can test only huge and complex theoretical systems rather than isolated hypotheses; but if we test two such systems which differ in one hypothesis only, and if we can design experiments which refute the first system while leaving the second very well corroborated, then we may be on reasonably safe ground if we attribute the failure of the first system to that hypothesis in which it differs from the other.

There are two problems with this idea. First and foremost, we can infrequently, if ever, implement this sort of strategy because the theoretical systems we use are indeed so 'huge and complex'. We are often interested, for instance, in the differences between theoretical systems like classical mechanics and special relativity! Second, moreover, the possibility of any crucial negative experiments must still be rejected because mixtures of true and false premises may be used to derive both true and false conclusions. In short, the first system may fail only because of the true hypothesis in which it differs from the second, which issues in a false prediction when combined with the rest of the system, and the second system may succeed only because of the false hypothesis in which it differs from the first, which issues in a true prediction when combined with the rest of the system.

An alternative strategy is therefore required. One suggests itself when we consider the following later passage:

> 'When do we—tentatively—accept a theory?' Our answer is, of course: 'When it has stood up to criticism, including the most severe tests we can design; and more especially when it has done this better than any competing theory.' (Popper 1983, p. 230)

My suggestion is that the answer to "When should we tentatively accept an auxiliary hypothesis?" is roughly the same when 'theory' is replaced by 'auxiliary hypothesis'. So what I propose, more particularly, is that the merit of an auxiliary hypothesis need be judged neither by its probability, e.g. $P(a,b')$ where b' is the part of b that goes beyond a, nor indeed just by 'how far it goes beyond the facts it is expected to explain'. Instead we may ask how well corroborated an auxiliary is, and so forth.

So if $C(h,e,b)$ is lower than the falsification threshold, one of our options is to examine the corroboration value of the auxiliaries in b, relative to tests of those auxiliaries that result in e_1:

$$C(A,e_1,b_1) = \frac{P(e_1,Ab_1) - P(e_1,b_1)}{P(e_1,Ab_1) - P(e_1A,b_1) + P(e_1,b_1)}$$

Note that b_1 does not, unlike b, contain A. (And yes, as the quick-witted will have realized, b_1 may contain auxiliaries, A_1, used to test A. A regress of testing is therefore possible, but I will come back to this.) So if we are interested in examining A, which we should be, we may consider how well it is corroborated *independently* of h. (And if it doesn't have a corroboration value at all, because it hasn't been tested, then we may take the opportunity to perform a test.) Furthermore, we may also split A into smaller components and examine the corroboration values of each of these (preferably each independently of h and of the other content in A). Ranking them with respect to corroboration value may then prove helpful in guiding our subsequent strategy. If one component of A turns out not to have a corroboration value, for example, then we should make it the target of a test forthwith (if we are able).

It is important to recognise that in performing such a ranking, we would usually take 'as fixed' many scientific theories other than h (which may, of course, be incorrect themselves). But in doing so, we need not (and should not) casually assume that those theories are true. Rather, we are in the business of systematically searching for mistakes; and we are starting, reasonably enough, with the (meta)hypothesis that the mistake is local. This is pragmatic, if nothing else. Why throw away the result of hard labour without seeing if it's any good?

One might nevertheless object that it is hard to see why starting in this way will generally be *best*. In fact, this objection is brought into stark relief by the recognition that corroboration value is not supposed to be any indication, whatsoever, of verisimilitude. We saw this in Chapter 2; see also Lakatos (1974) for a typical critique of Popper on this issue. As Popper puts it, in a passage added to *The Logic of Scientific Discovery* in 1972:

> The logical and methodological problem of induction is not insoluble, but my book offered a negative solution: (a) *We can never rationally justify a theory*, that is to say, our belief in the truth of a theory, or in its being probably true. This negative solution is compatible with the following positive solution, contained in the *rule of preferring* theories which are better corroborated than others: (b) *We can sometimes rationally justify the preference* for a theory in the light of its corroboration, that is, of the present state of the critical discussion of the competing theories, which are critically discussed and compared from the point of view of assessing their nearness to the truth (verisimilitude). The current state of this discussion may, in principle, be reported in the form of their degrees of corroboration.
>
> *The degree of corroboration is not, however, a measure of verisimilitude* . . . but only a report of what we have been able to ascertain

> up to a certain moment of time, about the comparative claims of the competing theories by judging the available reasons which have been proposed for and against their verisimilitude. (Popper 1999, pp. 281–282; emphasis added)

Bluntly, the objection would therefore be as follows. Why should we think there is something wrong with auxiliary a_1 rather than auxiliary a_2 simply because the corroboration value of a_2 is higher than that of a_1? (Of course, this question may appear to have even more force if the assumptions used to test the former are different from those used to test the latter. This will often be the case.) The only sensible answer I can see, within the confines of an anti-inductivist framework, is that we should not. Another reason is therefore required for advocating the methodological proposal that we explore auxiliaries with low (or no) corroboration values first (e.g. by subjecting them to further tests).

What could this be? In essence, I think it is as follows. By comparing the corroboration values of the auxiliaries (i.e. the components of A), and ranking them, we are identifying those areas where we have made the least (or even no) effort to check for errors.[9] Naturally, the fact that we have not checked for an error, in any given case, does not mean that there is an error in that case. (This is why we should not jump to the conclusion discussed in the previous paragraph.) To the extent that there might be an error there *and we have already checked all the other possibilities (more carefully)*, however, we should plausibly make some effort to look there next. Imagine, for example, that we were to find ourselves in the lucky position of knowing that either a_1 or a_2 is false. Imagine further that a_1 had been extensively tested, over the course of a hundred years, and had a corroboration value of close to unity, whereas a_2 lacked a corroboration value because it had never been tested. It would seem pretty natural to want to test a_2 (without assuming a_1), provided we were capable of doing so, rather than to continue to test a_1 *ad infinitum*. All that the critical rationalist adds to this, by her frank anti-inductivism, is the sombre truth that it does not follow, however unfortunately, that a_1 is more likely to be true than a_2. Rather, it shows that we haven't been able to find anything wrong with a_1 despite our best efforts, which we haven't yet directed elsewhere. There is an avenue we haven't explored. So perhaps we should explore it. Or to put matters somewhat differently, if we have no reason to expect a_1 to be true rather than a_2, and both are just as easy to test, then we don't have any good reason to devote all our effort to testing a_1. (We should also remember that in modern science, as we will see in the next chapter, this is a problem for groups rather than individuals. So we need not say that everyone working in the relevant area should perform the same tests, or refrain from working on the assumption that h has been falsified. Different individuals may rationally explore different paths; but the group as a whole may be constrained in such a way that "anything does not go"!)

Writing specifically of Bayesian confirmation, Strevens (2005, p. 914) suggests the following:

> 1. When *e* falsifies *ha*, the 'blame' is distributed between *h* and *a* roughly in proportion to their relative prior probabilities, so that a more probable *h* will be blamed relatively less.
>
> 2. The magnitude, positive or negative, of the impact of *e* on the main hypothesis *h* is greater the more probable the auxiliary *a*. When the probability of *a* is high, favourable evidence provides a greater boost to the probability of *h*, whereas unfavourable evidence makes a bigger dent in *h*'s probability.

In the case of corroboration, we may modify 2 to:

> 2*. When *e* falsifies (or corroborates) *ha*, the magnitude of the impact of *e* on the main hypothesis *h* is greater the more (independently) corroborated the auxiliary *a*.

A similar modification of 1 might also be suggested:

> 1*. When *e* falsifies *ha*, the 'blame' should be distributed between *h* and *a* roughly in proportion to their relative degrees of prior (independent) corroboration, so that a more corroborated *h* should be blamed relatively less.

Prima facie, it might seem that Popper would have objected to 1* on the basis that it serves to shield (main) hypotheses rather too much. But as intimated in the previous discussion, *a* and *h* are on precisely the same epistemological footing *qua* hypotheses (and it would be blithely uncritical to assume otherwise). As Popper (1959, p. 111) says in one of his most well-known passages:

> Science does not rest upon rock-bottom. The bold structure of its theories rises, as it were, above a swamp. It is like a building erected on piles. The piles are driven down from above into the swamp, but not down to any natural or 'given' base; and when we cease our attempts to drive our piles into a deeper layer, it is not because we have reached firm ground.

It should be noted, however, that although Strevens derives 1 and 2 from Bayes's theorem and the axioms of probability, it is not possible to derive 1* and 2* from the corroboration function in a similar fashion. For whereas C(*h*,*e*,*b*) is *defined* in terms of probability, it is not itself a probability. I am therefore proposing 1* and 2* on the basis that they are plausible in their own right. Nevertheless, I would not rule out the possibility that they might be derived from a more primitive *calculus of corroboration* (which is yet to be developed).

4. A FINAL OBJECTION

An excellent further objection, mentioned earlier in passing, is that we won't, generally, have really checked *all* the possibilities before we decide where to devote our attention. In examining the corroboration value of A, for example, we will require a further group of auxiliaries A_1. And if $C(A,e_1,b_1)$ falls below the falsification threshold, say, we may want to consider:

$$C(A_1,e_2,b_2) = \frac{P(e_2,A_1b_2) - P(e_2,b_2)}{P(e_2,A_1b_2) - P(e_2A_1,b_2) + P(e_2,b_2)}$$

In examining the corroboration values of that further group of auxiliaries, we will require yet another group of auxiliaries. And so on. Moreover, it cannot be the case that observation statements are a 'proper' (or privileged) terminus if these statements are themselves theory laden.

Here, it is appropriate simply to bite the bullet. Inquiry is a messy business, with no security and no guarantees, according to the fallibilist and deductivist framework of Popper (and subsequent philosophers falling broadly in the critical rationalist tradition). This is not so much how those of us who adopt this philosophical outlook *wish* that things were. It is, rather, how we think—we are sorry to say—that they are. So, yes, we may go on testing forever. And, yes, that process may lead us astray. But that isn't to deny that we should do our best to iron out our mistakes, given the resources at our disposal. I do believe, however, that it should lead us to reject the idea that truth is an aim (rather than a mere aspiration) of science. I will come to this a little later, in Chapter 7.

However, matters are not so dire as to mean that neither 1* nor 2*, discussed in the previous section, are ever applicable in practice. In order to see this, let's imagine that we have $C(h,e,b) \leq f$, where f is the falsification threshold we have agreed on. We want to know whether to blame h or A (which is a part of b), so we examine the corroboration of A *independent of h* (which is possible for some hypotheses and auxiliaries). We obtain $C(A,e_1,b_1)$, where b_1 does not contain h.

Let's now consider two possibilities:

(α) $C(A,e_1,b_1) > f$

(β) $C(A,e_1,b_1) \leq f$

In the event of α, we are in a position to apportion more blame to h than to A. And if $C(A,e_1,b_1)$ were much higher than f, then it would seem natural to blame h rather more than one would if $C(A,e_1,b_1)$ were only a little higher than f. Hence 2* appears to be applicable.

In the event of β, however, we want to know whether to blame A or A_1 (i.e. auxiliaries used in testing A, which are part of b_1). So we are not in a position to decide whether to blame h or A without further empirical investigation. We may therefore (independently) test A_1, if we can, and obtain $C(A_1,e_2,b_2)$. (Here we want b_2 to contain neither h nor A.) If this is above the falsification threshold, then we are in a position to blame A more than A_1. We are also, therefore, in a position to blame A more than h for the fact that our original $C(h,e,b) \leq f$. Only if $C(A_1,e_2,b_2) \leq f$ will we go on to worry whether A_1 or A_2 should be blamed. That's how an infinite regress of testing *might*, but by no means *must*, occur. Again, 2* appears to be applicable.

But what of 1*? Again, let $C(h,e,b) \leq f$. Now we need only imagine that the corroboration value of h was (independently) higher than that of A beforehand, or *vice versa*. In the former case, should we be more inclined to blame A? And in the latter, should we be more inclined to blame h? Provided both had corroboration values higher than f beforehand, this would be a natural reaction. Given the previously high corroboration of Newtonian mechanics (h), it appeared reasonable for Leverrier to posit Neptune (i.e. reject A) in the face of the anomalous orbit of Uranus (i.e. evidence that h and A was false) rather than question h. Similarly, it subsequently appeared reasonable to posit Vulcan in the face of the anomalous orbit of Mercury. The existence of such planets was, however, testable independently of h (at the time).[10] It is true that positing an invisible planet would have been seen as *ad hoc* in this instance, and would plausibly have been bad, but this situation is somewhat different to the one involving the neutrino, discussed earlier. To put it simply, this is because there were clearly *more easily testable* ways that A might have been false. In short, positing an invisible planet could only ever have been reasonable *after* those other testable options (not involving h) had been explored. But in the neutrino example, it was not clear how the hypothesis of energy conservation, despite its previous corroboration, might have been preserved *without* positing a particle such as a neutrino.

It is always possible that any of the propositions considered in a calculation of corroboration may be false—whether hypothesis, evidence, or background information. But this is a problem that confronts any account of confirmation or corroboration; a Bayesian can no more avoid the difficulty that background information may be false, and that inappropriate conditional probabilities might therefore be employed. The critical rationalist appears better off than the Bayesian here, because background knowledge is not taken as probably true, or anything like that. In the words of Popper (2002 [1963], p. 323):

> The fact that, as a rule, we are at any given moment taking a vast amount of traditional knowledge for granted (for almost all our knowledge is traditional) creates no difficulty for the falsificationist or the fallibilist. For he does not *accept* this background knowledge; neither as established nor as fairly certain, nor as yet probable. He knows that even its tentative acceptance is risky, and stresses that every bit of it is

> open to criticism, even though only in a piecemeal way. We can never be certain that we shall challenge the right bit; but since our quest is not for certainty, this does not matter.

The important lesson to take away is that there are decision-making resources at our disposal when we are faced with falsified conjunctions of statements on a corroboration-based view of scientific method, and that neither appeal to 'common sense' nor retreat to 'anything goes' is necessary.

ACKNOWLEDGEMENTS

This chapter is based on Rowbottom (In Press A).

6 The Roles of Criticism and Dogmatism in Science
A Group Level View

We have now covered corroboration and testing in a great deal of depth, and it is time to step back and look at the role that criticism can and should play in science as a whole. Even working on the assumption that criticism is important in some sense, which we will take as uncontroversial, we should question whether it is correct to suggest that each and every scientist should have a critical approach, or whether it is enough for science to be organised so as to perform a critical function. We may also ask whether, and to what extent, dogmatism is important in science. In order to do so, it seems fitting to revisit a classic debate between Kuhn and Popper.

1. CRITICISM AND THE GROWTH OF KNOWLEDGE

Criticism and the Growth of Knowledge was the flashpoint for a well-known debate between Kuhn and Popper, in which the former emphasised the importance of 'normal science' *qua* puzzle solving and the latter (and his critical rationalist supporters) questioned the very idea that 'normal science', so construed, could count as good science at all.

Kuhn's basic idea was that science would hardly get anywhere if scientists busied themselves with attacking what they already had, rather than accepting and refining it. But as we have seen, Popper instead emphasised the importance of striving to overthrow theories that might very well, for all their designers and users knew, be false (or otherwise unfit for purpose). On the face of it, both had perfectly reasonable points.

The natural solution would have been to take the middle ground—i.e. to suggest that it is acceptable for some scientists to be dogmatists and for others to be highly critical—but both Popper and Kuhn appear to have avoided this option because they ultimately considered the matter, I will contend, only at the level of the individual scientist. By this, I mean that each thought about how a lone scientist ought to behave, and then extrapolated from this to determine how they thought a group of scientists should behave. And, on the one hand, it is hard to see how it can be right for an individual (and by extrapolation every) scientist to blithely assume that what everyone else is doing is basically right, and that the theories she is given are fit for purpose. On the other, it is difficult to see how it could be right for a (and by extrapolation every) scientist to treat the canon of her science as fundamentally wrong, and to spend all her days objecting to the basic metaphysical principles underlying it.

In this chapter, I will explain how thinking at the level of the group, using a functionalist picture of science, provides a means by which to resolve the tension. I do not claim that this notion is entirely new—Hull (1988), Kitcher (1990), and Strevens (2003) all discuss the importance of the division of cognitive labour (although not with emphasis on attitudes), and Jarvie (2001, p. 5) claims that Popper himself argued 'that traditional problems of epistemology need to be reformulated socially'[1]—but the subsequent analysis is considerably more penetrating than that that recently offered by Domondon (2009) and Fuller (2003). In fact, it shows precisely how we can resolve, and move beyond, the kind of 'Kuhn versus Popper' problems with which these authors are concerned.

2. POPPER ON CRITICISM

As we have already seen, Popper repeatedly pushed the idea that a critical attitude is at the heart of the scientific persona, and that a critical method is its proper counterpart. Despite his well-known emphasis on the importance of falsifiability, he acknowledged even in the original version of *The Logic of Scientific Discovery* (i.e. in 1934 but only translated into English in 1959) that:

> A system such as classical mechanics may be 'scientific' to any degree you like; but those who uphold it dogmatically—believing, perhaps, that it is their business to defend such a successful system against criticism as long as it is not *conclusively disproved*—are adopting the very reverse of that critical attitude which in my view is the proper one for the scientist. (Popper 1959, p. 50)

This theme runs through his work before the 1970s—e.g. Popper (1940, p. 404), Popper (2003 [1945], vol. 2, p. 249), Popper (1959, p. 16), Popper (2002 [1963], p. 67), and Popper (1968, p. 94)—which throws the following quotation, in *Criticism and the Growth of Knowledge*, into rather sharp relief:

> I believe that science is essentially critical . . . But I have always stressed the need for some dogmatism: the dogmatic scientist has an important role to play. If we give in to criticism too easily, we shall never find out where the real power of our theories lies. (Popper 1970, p. 55)

Only at two points before 1970, however, did Popper suggest that there is a need for dogmatism in science. One such passage is as follows:

> [D]ogmatism allows us to approach a good theory in stages, by way of approximations: if we accept defeat too easily, we may prevent ourselves from finding that we were very nearly right. (2002 [1963], p. 64)[2]

However, the extent to which dogmatism is useful, according to this view, is only in so far as we are fallible. That is to say, dogmatism will only prove useful on those occasions where we are 'very nearly right' despite evidence to the contrary. But how about when we are wrong? And, furthermore, might we not accept a methodological rule such as 'Do not accept that a theory is falsified too easily' (as suggested in the previous chapter) without being dogmatic at all? We will return to these questions when we have looked at what Kuhn had to say about dogmatism.

3. KUHN ON DOGMATISM

In contradistinction to Popper, Kuhn suggested that adherence to the *status quo* was characteristic of actual 'normal', and derivatively good, science.[3] Infamously, Kuhn (1996 [1962], p. 80) claimed that an experiment which backfires is normally taken, *and should normally be taken*, to reflect badly on the scientist that performs it: 'Failure to achieve a solution discredits only the scientist and not the theory . . . "It is a poor carpenter who blames his tools" . . . '

Moreover, Kuhn (ibid., p. 80) suggested that normal science can enable us 'to solve a puzzle for whose very existence the validity of the paradigm must be assumed'. So, in short, he thought that work within a paradigm (*qua* disciplinary matrix) is possible only if that paradigm is taken for granted. Later in *The Structure of Scientific Revolutions*, he expressed this view at greater length:

> [T]rial attempts [to solve puzzles], whether by the chess player or by the scientist, are trials only of themselves, not of the rules of the game. They are possible only so long as the paradigm itself is taken for granted. (Kuhn 1996 [1962], pp. 144–145)

As I have argued elsewhere (Rowbottom 2006b), this is an elementary mistake. On the contrary, the puzzles 'exist' *whether or not* the paradigm is assumed to be valid (or taken for granted), because *we only need to consider what would be the case if the paradigm were valid (or entertain the paradigm) in order to examine them.* By way of analogy, we can consider whether p & q follows from p and q in classical logic without assuming that both *p* and *q* are true, or indeed assuming that classical logic is fit for purpose (whatever our stance on psychologism in the philosophy of logic happens to be).

Kuhn (1996 [1962], p. 24) used similarly strong language in other passages:

> By focusing attention upon a small range of relatively esoteric problems, the paradigm *forces* scientists to investigate some part of nature in a detail and depth that would otherwise be unimaginable . . . (my emphasis)

However, we might interpret this a little more loosely by thinking about what is 'forced' in order to demonstrate one's proficiency in, and even to remain a recognised worker in, a discipline. In essence, Kuhn appears to have thought that scientists would not be *motivated* to tackle such esoteric problems (or puzzles) without rigid belief—or even faith—in the paradigm.[4]

This claim may also be somewhat dubious, however, because it is possible to do things for extrinsic reasons. I could learn to recite a poem in order to impress a prospective lover without having any interest in verse or metre, just as a scientist could be content to solve puzzles, in the short term, in order to support himself and slowly build a reputation which would lead to some of his potentially revolutionary ideas being taken more seriously by his peers.

Nevertheless, Kuhn had a valid point to the extent that he was worried about scientists becoming *hypercritical*: 'The scientist who pauses to examine every anomaly he notes will seldom get significant work done.' (Kuhn 1996 [1962], p. 82) And this, as we will see in the next section, is where Kuhn's objection to Popper's emphasis on criticism may be thought to have some bite. It won't do to have everyone criticise whatever they like, willy-nilly.

4. KUHN VS. POPPER IN CRITICISM AND THE GROWTH OF KNOWLEDGE

The stage is now set for an examination of the exchange between Popper and Kuhn in *Criticism and the Growth of Knowledge*. We have seen that the former emphasised the importance of criticism and non-conformity in science, whereas the latter thought that conformity and focused puzzle solving are essential (at least in 'normal science'). It is therefore clear that they were set on a collision course when they were brought together, as Gattei (2008, ch. 2) illustrates.[5]

Perhaps we should start by noting that Popper agreed with Kuhn that 'normal science' exists. Unsurprisingly, however, he did not describe it in flattering terms:

> "Normal" science, in Kuhn's sense, exists. It is the activity of the non-revolutionary, or more precisely, not-too-critical professional: of the science student who accepts the ruling dogma of the day; who does not wish to challenge it; and who accepts a new revolutionary theory only if almost everybody else is ready to accept it—if it becomes fashionable by a kind of bandwagon effect. (1970, p. 52)

Popper (ibid.) continued by expressing pity for the predicament of 'normal scientists', with reference to educational norms: 'In my view the "normal" scientist, as Kuhn describes him, is a person one ought to be sorry for . . . He has been taught in a dogmatic spirit: he is a victim of indoctrination. He has learned a technique which can be applied without asking for the reason why (especially in quantum mechanics) . . . '

Kuhn disagreed with Popper not because he thought that criticism is unimportant for scientific progress (whatever that may consist in[6]), but rather because he thought that it should only be occasional. (We can admit, of course, that puzzle solving involves *some* criticism. The point is just that this has narrow scope.) Kuhn summarised his view as follows:

> Sir Karl . . . and his group argue that the scientist should try at all times to be a critic and a proliferator of alternate theories. I urge the desirability of an alternate strategy which reserves such behaviour for special occasions . . . Even given a theory which permits normal science . . . scientists need not engage the puzzles it supplies. They could instead behave as practitioners of the proto-sciences must; they could, that is, seek potential weak spots, of which there are always large numbers, and endeavour to erect alternate theories around them. Most of my present critics believe they should do so. I disagree but exclusively on strategic grounds . . . (Kuhn 1970b, pp. 243, 246)

But what are the strategic grounds upon which Kuhn made his recommendation to reserve criticism of theories for special occasions? His fundamental idea was that only by working positively with our current theories for a considerable period—trying to refine them, improve and increase their applicability, and so forth[7]—can we discover their true strengths and weaknesses. So when we do decide that change is needed, we will know where to focus our attention:

> Because that exploration will ultimately isolate severe trouble spots, they [i.e. normal scientists] can be confident that the pursuit of normal science will inform them when and where they can most usefully become Popperian critics. (Ibid., p. 247)

One problem with Kuhn's suggestion is that he leaves it so vague. It is not clear, for instance, what counts as a severe trouble spot (and who should get to decide). Furthermore, it is unclear how long we should stick with a theory in the face of trouble spots. And finally, crucially, it is unclear why working with a theory for a long time should improve the chance of isolating *genuine* limitations of the theory. This is evident when we consider Kuhn's proposed strategy in the light of Duhem's thesis, which we covered in the previous chapter.

In short, the salient question is again "When should one challenge the theory itself, rather than the auxiliary assumptions used in order to derive predictions from it?" Kuhn seems to have suggested that the auxiliaries do (and should) always give way in 'normal science'.[8] Naturally this is completely at odds with Popper's (1959, p. 83) dictum that: 'As regards *auxiliary hypotheses* . . . only those are acceptable whose introduction does not diminish the degree of falsifiability or testability of the system in question.' It is also at odds with the treatment of corroboration in the previous chapter, where we saw that it is possible to consider the corroboration of auxiliaries in order to inform decision making.

With this in mind, let us now revisit the passage mentioned at the beginning of the chapter:

> [T]he dogmatic scientist has an important role to play. If we give in to criticism too easily, we shall never find out where the real power of our theories lies. (Popper 1970, p. 55)

Popper might instead have said that we should be willing to *criticise* auxiliary hypotheses as well as theories, and that we shouldn't be too quick to condemn the latter rather than the former. But there is quite a difference between saying this and saying that 'the dogmatic scientist has an important role to play', because it is possible to attack auxiliary hypotheses rather than theories, in the light of evidence that falsifies the conjunction thereof, *without* being dogmatically committed to the theories. Therefore Popper did not intend 'dogmatism' in the same sense that Kuhn did, as he went on to point out:

> [T]his kind of dogmatism is not what Kuhn wants. He believes in the domination of a ruling dogma over considerable periods; and he does not believe that the method of science is, normally, that of bold conjectures and criticism. (Ibid.)[9]

Popper's comment here is fair, because Kuhn was much more extreme in his claims about the value of shielding theories from criticism. The following two quotations, in particular, illustrate this:

> It is precisely the abandonment of critical discourse that marks the transition to a science. (Kuhn 1970a, p. 5)

> Lifelong resistance, particularly from those whose productive careers have committed them to an older tradition . . . is not a violation of scientific standards. (Kuhn 1996 [1962], p. 151)

It is not entirely misleading, therefore, to paint Popper and Kuhn as two extremists on the issue of the role of criticism in (ideal) science. On the one hand, Popper suggested—at many points in his writing, at least—that:

(P) Each and every scientist should have a critical attitude (and follow the same critical procedures).

On the other, Kuhn—or at least a slight caricature of Kuhn[10]—suggested that:

(K) Each and every scientist should puzzle solve within the boundaries of the disciplinary matrix, on the basis of the exemplars therein, until almost every scientist comes to see particular failures as indicating serious anomalies.

I should emphasise that (K) only goes for 'normal science', and that 'puzzle solving' involves many different forms of activity (as shown later in Figure 6.5).[11] It is crucial to the plausibility of Kuhn's view that 'rational' disagreement between scientists (Kuhn 1977, p. 332) *is permissible during periods of extraordinary science.*

According to either (P) or (K), each and every scientist is expected to perform the same functions, *qua* scientist.[12] Failure to do so will lead to something less than ideal science. In what follows, I shall challenge this notion that all scientists should adopt similar stances, and suggest that the best possible science may be realized in more than one way. I will also suggest that there is a place for dogmatism in something close to Kuhn's sense when we look at matters at the group level, but that critical procedures are also crucial. The key idea, as Kitcher (1990, p. 6) puts it, is that there is 'a mismatch between the demands of individual rationality and those of collective (or community) rationality'.

It should be noted, however, that the question of how we should divide cognitive labour arises even if the function that we should all perform is identical. Imagine, for instance, that we should all be critical. It does not follow that we should all start from the same assumptions, or work on the same research programmes. If the relative progression of programme P is greater than that of programme R, then perhaps more effort should be devoted to P than to R. Abandoning R altogether may be irrational at the group level. Kitcher (1990) provides several examples to this effect.

So, in short, the problem of the division of cognitive labour exists independently of the problem here discussed. To see this, imagine that the problem to be worked on (and relevant minimal set of assumptions) is fixed. What kind of researchers do we want to use? Do we want lots of highly critical gals? Do we want lots of dogmatic guys? Or do we want a balance, perhaps because highly critical gals are better suited to performing some functions than dogmatic guys, and *vice versa*? Maybe it's even possible that critical gals and dogmatic guys can achieve things when working together that they could never achieve when working apart.

5. A FUNCTIONAL ANALYSIS

The differences between Kuhn and Popper can be neatly understood by thinking in terms of functions, as I will show in the following. Moreover, thinking in this way suggests a means by which to resolve their debate; namely to consider functions at the group, rather than the individual, level.

The simple Popperian scientist fulfils two functions (which fall inside the grey area in Figure 6.1): one imaginative, and the other critical.[13] In short, the scientist uses propositions from outside sources (e.g. tradition and experience) to criticise his hypotheses (which are often derived from his imagination). The critical function may involve several procedures, e.g. non-empirical checks for internal consistency as well as empirical tests. Those hypotheses that survive the process count as corroborated (and are outputs). But simply because a hypothesis is corroborated, this does not mean that it is no longer subject to the critical function (and hence the bidirectional arrow between 'critical' and 'corroborated hypotheses').

There is, however, one rather striking feature of the Popperian scientist so depicted; he is purely theoretical in orientation. This appears to be the

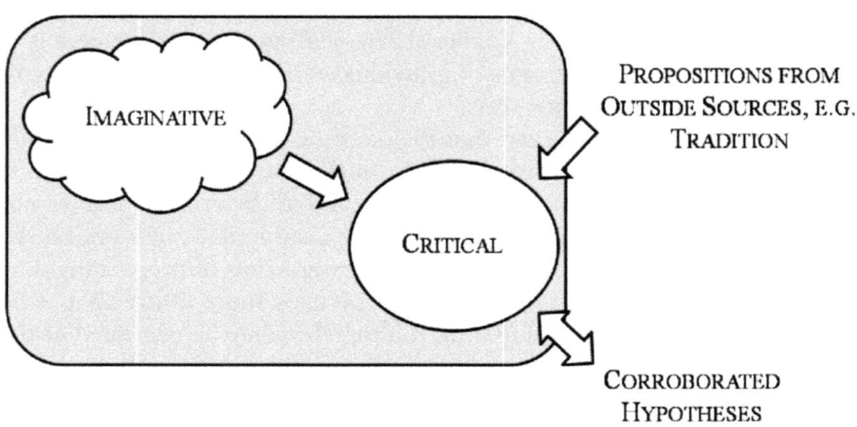

Figure 6.1 The simple Popperian scientist.

correct view of Popper's position because he suggests that applied science is the province of 'normal scientists':

> [The normal scientist] has become what may be called an applied scientist, in contradistinction to what I should call a pure scientist. He is, as Kuhn puts it, content to solve 'puzzles' . . . it is not really a fundamental problem which the 'normal' scientist is prepared to tackle: it is, rather, a routine problem, a problem of applying what one has learned . . . (Popper 1970, p. 53)

So what should we think of Popper's mention of dogmatism? If we imagine this as a function at the individual level, we will arrive at a somewhat different view from that depicted in Figure 6.1. Instead there will be a 'dogmatic filtering' function, in addition to the critical and imaginative functions, which will serve to ensure that some propositions—and in particular, some theories—are not criticised. As we can see from Figure 6.2, however, such a filter need not be 'dogmatic' in any strong sense of the word. This is because the filter may function such that no (empirical) theory is *in principle* immune to being passed on to the critical procedure. So if a theory is brand-new, for instance, perhaps it will be shielded from criticism until it can be further developed (by the imaginative, or creative, function); hence the bidirectional arrow between the imaginative and filtering functions. But if a theory is well developed, i.e. has had a lot of imaginative effort spent on it, perhaps it will always pass through the filter.

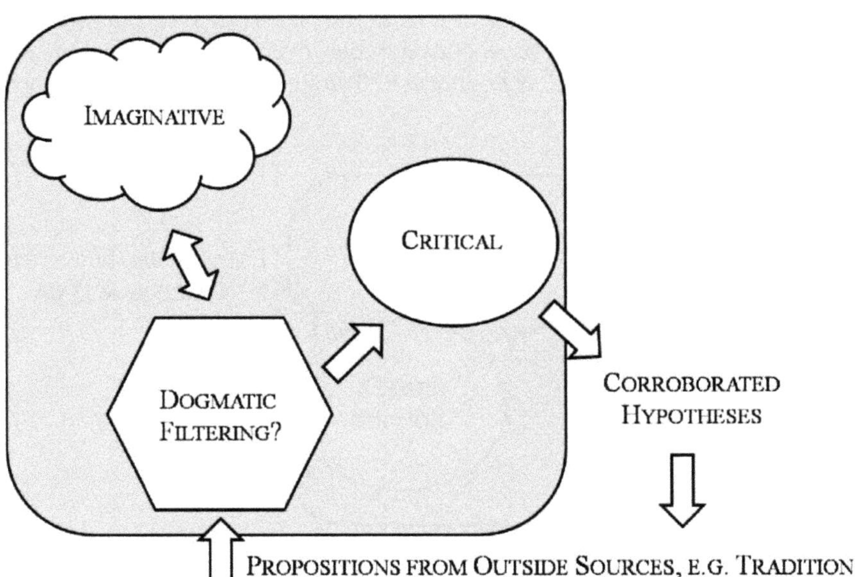

Figure 6.2 The sophisticated Popperian scientist.

Let us now compare this with the Kuhnian normal scientist. In contrast to her Popperian counterpart, her primary function is to solve puzzles. And in order to do this, she relies on established scientific theories and data. (We should allow that some of the data used may not itself be a product of science. However, in so far as observations are heavily theory laden, on Kuhn's view, it is likely that said data will be given an interpretative slant—and/or that what counts as admissible data will be determined—by the disciplinary matrix.) The outputs of puzzle solving are both theoretical *and* concrete; that is to say, Kuhn does not draw a sharp distinction between 'pure' and 'applied' science in the manner that Popper does.

We might wonder, though, whether good puzzle solving requires imagination, and therefore if the imaginative function is not also, as depicted in Figure 6.3, a required component of the Kuhnian scientist. Despite first appearances, a somewhat closer look at Kuhn's position appears to suggest that it is not, because exemplars provide *templates* for puzzle solving. As Bird (2004) puts it:

> In the research tradition it inaugurates, a paradigm-as-exemplar fulfils three functions:
> (i) it suggests new puzzles;
> (ii) it suggests approaches to solving those puzzles;
> (iii) it is the standard by which the quality of a proposed puzzle-solution can be measured.[14]

To remove the 'imaginative' function from the picture is not to suggest that puzzle solving does not require considerable ingenuity, on occasion, nor indeed that it is as 'routine' as Popper (1970) suggested. The point is simply

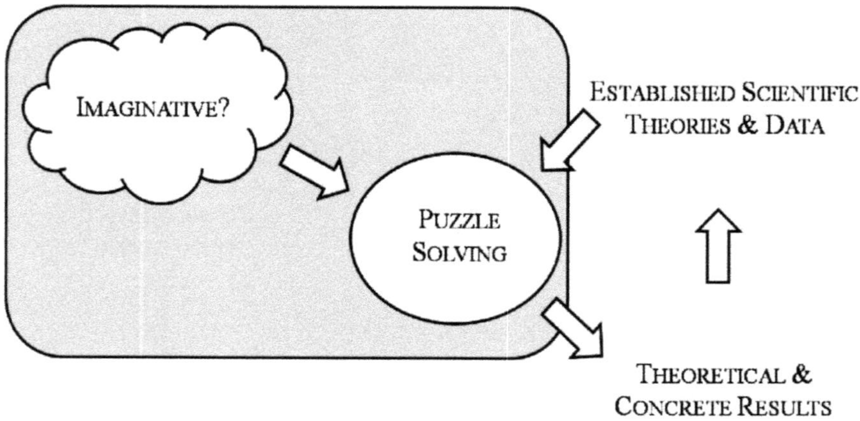

Figure 6.3 The *prima facie* Kuhnian normal scientist.

that an incredibly difficult puzzle is still little more than a puzzle when the rules of the game and procedures for playing are all fixed.[15] And Kuhn certainly does not suggest that (normal) scientists require anything like '"an irrational element", or a "creative intuition", in Bergson's sense' (Popper 1959, p. 32). On the contrary:

> The paradigm he has acquired through prior training provides him [i.e. the normal scientist] with the rules of the game, describes the pieces with which it must be played, and indicates the nature of the required outcome. His task is to manipulate those pieces within the rules in such a way that the required outcome is produced. (Kuhn 1961, p. 362)

We are therefore left with the picture that follows, depicted in Figure 6.4, in which exemplars remove the need for an imaginative function. (It is worth adding that an imaginative function may be required in extraordinary science, e.g. in order to bring exemplars into being, but that we are not presently concerned with this.)

Figure 6.4 does run the risk, however, of making Kuhn's picture look rather simpler than it actually is.[16] This is because many different activities fall under the rubric of 'puzzle solving', as Kuhn explains in Chapter 3 of *The Structure of Scientific Revolutions*. So Figure 6.5 gives a look inside the puzzle solving function, and shows that it is composed of several different processes. For a full discussion of these processes—classification and prediction, theory-experiment alignment, and articulation—see Rowbottom (In Press D). For present purposes, it suffices to note that these are significant functions within the function of puzzle solving.

We have now seen that despite their strikingly different views of the ideal scientist, both Popper and Kuhn had understandings that can be modelled with ease via a functional perspective. For both, to be a good scientist is

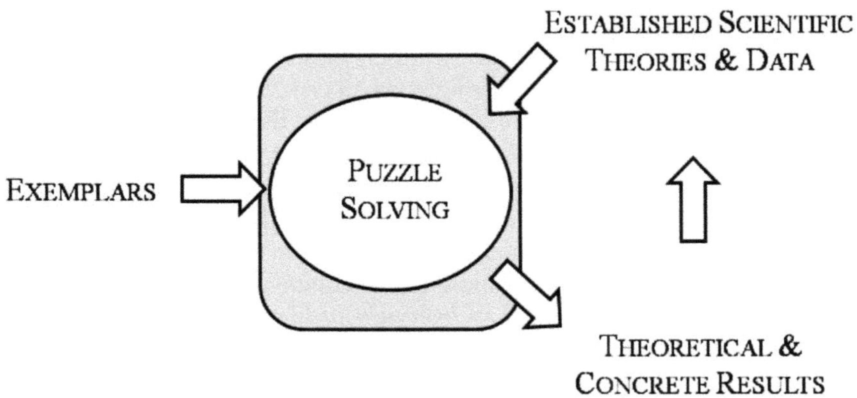

Figure 6.4 The Kuhnian normal scientist.

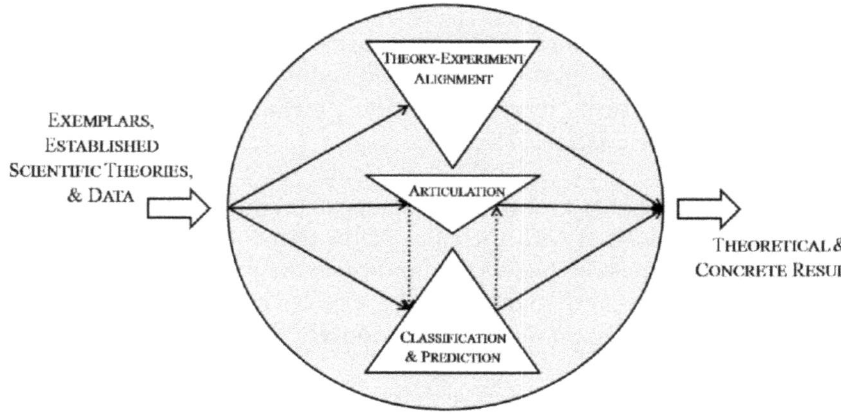

Figure 6.5 Inside puzzle solving.

simply to perform specific functions. And good science is to be understood as an activity performed by large numbers of good scientists in precisely the aforementioned sense.

However, this functional analysis makes the following questions, which we will come to in the next section, salient. Why not have different functions performed by different scientists? And why not entertain the possibility that it is (sometimes) *necessary* for the functions to be performed by different individuals in order for science to be (or to be as close as possible to) ideal?

6. FUNCTIONS AT THE GROUP LEVEL: A HYBRID MODEL

Moving to a consideration of functions at the group level allows us to consider the possibility that both dogmatism and criticism are vital components of the scientific enterprise. And while it may be suggested that Kuhn would have agreed in so far as criticism might play a crucial part in extraordinary science, his picture is one where science *should* go through different phases. In short, his view appears to have been that either all scientists should be doing normal science, or all scientists should be doing extraordinary science. The possibility that it might be preferable for the two kinds of activity to co-exist, in so far as some might be dogmatic and others might be critical at the same time, was never dismissed on adequate grounds.

Allow me to start by giving an overview of Figure 6.6, which may initially appear impenetrable. According to this model, (ideal) science involves each of the three primary functions discussed previously: imaginative, puzzle solving, and critical. The imaginative function provides *some* objects of criticism, which may be evaluated and rejected, or defended, attacked,

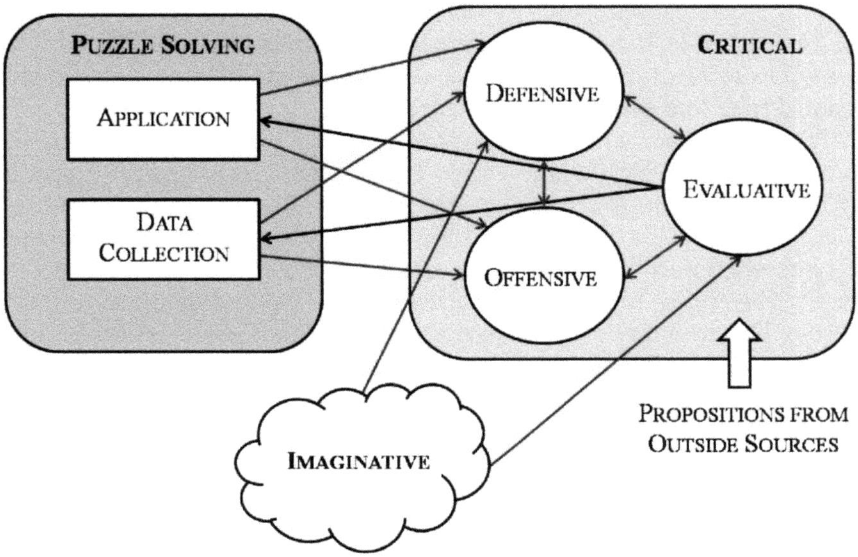

Figure 6.6 A hybrid view of science at the group level.

and then subsequently evaluated. (Note also that said evaluation may rely on propositions from outside sources, e.g. tradition, too.) The critical function has three parts: offensive, defensive, and evaluative. These should be reasonably self-explanatory, but will be illustrated in the course of the subsequent discussion.

Now it is crucial to distinguish between critical procedures (or methods) and the critical attitude. That is to say, *it is possible for science to perform a critical function with wide scope even when none of its participants have (completely) critical attitudes.* One simple way to see this is to imagine a scenario in which each scientist holds different assumptions dogmatically, but in which no peculiar assumption (*qua* proposition or theory) is held dogmatically by all scientists. *So at the group level no statement is beyond criticism.*

We can develop this idea by considering another simple scenario in somewhat greater depth. Imagine two dogmatists, D_1 and D_2, who are dogmatic only in so far as they will do anything to defend their individual pet theories, T_1 and T_2, which are mutually exclusive. So D_1 will *defend* T_1 at all costs, e.g. by challenging auxiliary statements used to generate predictions from T_1 in the event of the possibility of empirical refutation, just as D_2 will defend T_2. (This is fulfilling a *defensive* function.) To the death, neither D_1 nor D_2 will abandon their respective pets. But each will try to persuade other scientists—whether or not they try to persuade one another—that their own pet is superior. And as part of this, said dogmatists need not only defend their own pets against attack, but may also attack the rival pets of others. So part of D_1's strategy to promote T_1 may be to attack T_2, just as part of D_2's strategy to promote T_2 may be to attack T_1 (i.e. to fulfil an *offensive* function). Thus both dogmatists may fulfil (narrowly focused) critical functions of attack and defence.[17]

Yet if everyone were such a dogmatist, stalemate (and perhaps even disintegration of science) would ensue. This is why a third critical function, that of evaluation, is crucial in order to judge whether T_1 or T_2 emerges victorious. Needless to say, such an evaluative function may be performed by interested third parties who are not themselves committed to either T_1 or T_2. Therefore a third dogmatic scientist, D_3, who is set upon defending a theory T_3, which may stand irrespective of whether T_1 or T_2 is correct, may serve as an evaluator of the debate between D_1 and D_2. In short, to attack or defend as a result of dogmatism in one context does not preclude evaluating in another.

But how might dogmatic individuals benefit science in a way that their completely critical (and/or highly evaluative) counterparts might not? The simple answer is that they may be far more persistent in defending their pet theories (and therefore attacking competitor theories) than a more critical individual could be. So they might, for example, consider rejecting auxiliary hypotheses when their critical counterparts would not (and would instead simply reject a theory). But it is worth reiterating that being dogmatic in this sort of sense does not preclude being critical. *Rather, the critical activity of such an individual will have narrow scope; it will be aimed only at defending pet theories and attacking competitor theories.* So, in short, to be critical in some small area is still to be critical, even though it is not necessarily to have the critical attitude that '*I may be wrong and you may be right, and by an effort, we may get nearer to the truth*' (Popper 2003 [1945], vol. 2, p. 249), or to be a pancritical rationalist in the sense of Bartley (1984) as discussed in Chapter 1. Just as there are occasions where 'a commitment to the paradigm was needed simply to provide adequate motivation' (Kuhn 1961, p. 362), there may be occasions where dogmatic commitment is crucial in order to push the scientist to consider avenues that would be ruled out by more open-minded, evaluative individuals. My point here is that territory may be explored which would otherwise not be, and that this might result in a variety of fruits; I do not join Kuhn (1996 [1962], p. 247) in thinking that such exploration will, *as a general rule*, be successful in isolating 'severe trouble spots'.

Furthermore, it may be a good thing for individual scientists to devote themselves to performing a small number of functions.[18] Perhaps, for instance, it is extremely difficult (due to human limitations) to be an expert puzzle solver and an expert attacker. Perhaps, indeed, the kind of person who is an expert attacker is often a lousy puzzle solver (because he or she finds it hard, *qua* boring, to work with externally imposed frameworks of thought or to perform repetitive tasks). So here we might again, as in Chapter 1, say that something like van Fraassen's (2002,

2004a) notion of a stance is relevant, especially if we think of this as involving a *mode of engagement* and a *style of reasoning* (Rowbottom and Bueno, In Press A). One scientist's style may be to think outside the box, and he might engage by performing wild new experiments or working on highly abstract theories. Another scientist's style may be to think inside the box, and she may engage by repeating well-known experiments (with minor refinements). Trying to force either scientist to change style or mode may be unwise. Indeed, it may not always be possible for such changes to occur.[19]

In closing this section, we should also consider how the puzzle solving function may relate to the critical one. (Consider, again, Figure 6.6.) First, only theories which are positively evaluated (by those performing the evaluative function) will be used for puzzle solving purposes. It is these theories that will be applied, and which will determine what sort of data is normally considered to be worthy of collection. Second, however, the puzzle solvers' data and results may be useful to those performing critical functions of attack and defence. (Attempts to puzzle solve may isolate unanticipated trouble spots, for example, just as Kuhn suggested.) Third, the results of the attack and defence processes will be evaluated, and this will determine what sort of puzzle solving takes place next. So, in short, there may a fruitful interchange between puzzle solvers and criticisers; and perhaps this is the genuine lifeblood of science.

7. FURTHER QUESTIONS

The model proposed here raises quite different questions from those explicitly tackled by Kuhn and Popper, and shifts the focus of the debate. How should the balance between functions be struck? That is to say, for any given group of scientists, how many should be fulfilling puzzle solving functions, rather than critical functions? And of those performing critical functions, how many should, say, be performing evaluative functions? These questions, and others like them, plausibly do not have 'hard and fast' contextually invariant answers. Instead, the proper distribution of activity may depend on the skill base available—e.g. perhaps despite their best efforts, some people cannot feign being dogmatic when they are not, in so far as they cannot really push themselves to defend a theory come what may—and also the state of science at the time. If T_4 were evaluated as suffering from severe defects but there was nothing else available to put in its place, for instance, then perhaps more imaginative effort would be required. Similarly, if T_5 remained untested and unchallenged, then perhaps more offensive and defensive interplay concerning T_5 would be merited. (So note also that the wisdom of occasional episodes resembling revolutions, but not quite so extreme and wide-ranging as Kuhn's model demands, may be accounted for.)

I should emphasise that I have not denied that there is a fact of the matter about what an individual scientist might best do (or best be directed to do) in a particular context of inquiry. Rather, I have suggested that determining what this is requires reference not only to the state of science understood as a body of propositions (or as knowledge) but also to what other scientists are doing and the capacities of the individual scientist. (In short, we can make sense of the question by holding some variables fixed.) Consider a new professional scientist, going into his first postdoctoral research project; and let his capacity for good work be fixed by his interests, desires, and experience. Assume he could work just as well in group B as in group A. It might be preferable for him to join the latter because its line of inquiry is more promising than that of the former, on current indicators, although it has fewer members. So, in short, I take there to be measures—even if they are rough measures, such as Popper's corroboration function—of how theories (and/or research programmes, modelling procedures, etc.) are faring. And these, given the resources at our disposal, determine how we should respond.

A simple analogy may help. Imagine you, the chess player, are managing science. The pieces are the scientists under your command, and their capacities vary in accordance with their type (e.g. pawn or rook). The position on the board—nature is playing the opposing side—reflects the status quo. And now imagine you are told that, against the rules of normal chess, you are allowed to introduce a new pawn (which you can place on any unoccupied square).[20] (This is akin to the introduction of a new scientist; pieces working in combination on your side can be thought of as research groups, and so on.) Some moves will be better than others, given your aim of winning the game, and in some circumstances it will be clear that one available move is best.

So my own view is that considering social structure neither precludes employing insights from what might be called the 'logical' tradition in the philosophy of science—formal apparatus, such as corroboration functions, for instance—nor requires acceptance of the view that studies in scientific method always require reference to the history of science. Social structure is relevant to questions of scientific method; but it is hardly as if when we discuss groups, rather than individuals, we suddenly find ourselves in territory where the 'logical' tradition has nothing to offer.[21]

The picture presented in this chapter is complex, and the questions enumerated in this concluding section are daunting. It may prove to be the case that they are beyond our power to answer satisfactorily except in idealized contexts. Nevertheless, it appears that complexity is necessary if we are to truly get to grips with the question of how science should work. At the very least, the model here considered, e.g. as presented in Figure 6.6, provides a basic framework with which to tackle practical questions when considering the research activity of a group (or groups). And even if that model is rejected, to consider functions *at the level of the group* is to make an important conceptual breakthrough in understanding (and therefore shaping) science.

What does all this tell us about critical rationalism? First, it shows that while it may indeed be applicable at the level of the individual, it is plausibly not applicable at the level of the group. To be an ideal individual inquirer, it may be necessary to be a pancritical rationalist (or as close as one can be). But for group inquiry to be ideal, it may often be necessary for it to involve individuals that are not themselves ideal inquirers. Second, however, this much of critical rationalism remains even at the group level. *It is crucial for ideal group inquiry to involve critical functions, and these should lie at its very core.* In summary, it may be a good thing for science that some scientists are dogmatists. But without scientists performing critical functions (such as attack and defence of theories)—and not just in sporadic 'extraordinary' periods—science would simply cease to exist.

ACKNOWLEDGEMENTS

This chapter is based on Rowbottom (In Press C).

7 The Aim of Science and Its Evolution

Popper was a realist, and thought that the aim of science is to find true, or increasingly truthlike, theories. But the key idea behind his methodological proposals was that we should proceed by ruling out false theories. Popper employed an evolutionary analogy—or so-called 'evolutionary epistemology'—in an attempt to bridge the gap and argue that this means can achieve the aim.

After illustrating what it takes for something to count as the aim of science, this chapter explores what might count as such an aim on an evolutionary analogy. It shows that even if our observations and experimental results are reliable, an evolutionary analogy fails to demonstrate why conjecture and refutation should result in: (1) the isolation of true theories; (2) successive generations of theories of increasing truth-likeness; (3) empirically adequate theories; or (4) successive generations of theories of increasing proximity to empirical adequacy (or even structural adequacy). It concludes that an evolutionary analogy is only sufficient to defend the notion that the aim of science is to isolate a particular class of false theories, namely, those that are empirically inadequate. The upshot is that (pan) critical rationalists should accept that this is the primary aim of science, given that the evolutionary analogy is apt.

1. EVOLUTION AND SCIENTIFIC PROGRESS

The idea that scientific progress may be explained by analogy with evolution, and with natural selection in particular, is now commonplace. In fact, such 'evolutionary epistemology' may be described as a 'program [which] attempts to account for the evolution of ideas, scientific theories and culture in general by using models and metaphors drawn from evolutionary biology' (Bradie 1990, p. 246).[1] Key twentieth-century advocates of this programme are Toulmin (1967), Popper (1959, 1972, 1984), Campbell (1974), Lorenz (1977), and Hull (1988). Moreover, other philosophers of science often appeal to evolutionary analogies at crucial points in their discussions.

Van Fraassen (1980, p. 40), for example, supports his constructive empiricist view as follows:

> [T]he Darwinist says: Do not ask why the *mouse* runs from its enemy. Species which did not cope with their natural enemies no longer exist. That is why there are only ones who do. In the same way, I claim that the success of current scientific theories is no miracle . . . For any scientific theory is born into a life of fierce competition, a jungle red in tooth and claw. Only the successful theories survive—the ones which *in fact* latched on to actual regularities in nature.

An early example of the use of the evolutionary analogy by Popper (1959, p. 42), long before 'evolutionary epistemology' was ever mentioned, is:

> [W]hat characterizes the empirical method is its manner of exposing to falsification, in every conceivable way, the system to be tested. Its aim is not to save the lives of untenable systems but, on the contrary, to select the one which is by comparison the fittest, by exposing them all to the fiercest struggle for survival.

In this chapter, we will take a fresh look at the extent to which such analogies are appropriate for describing scientific progress, and consider in particular how we might understand the proper aim of science in light of possible and actual selection processes. Is it truth, empirical adequacy, or something altogether different? We will work on the assumption that the logic of selection is always the same. As Okasha (2006, p. 10) puts it:

> [I]t is easy to see that Darwin's reasoning applies not just to individual organisms. *Any* entities which vary, reproduce differentially as a result, and beget offspring that are similar to them, could in principle be subject to Darwinian evolution. The basic logic of natural [and presumably non-natural] selection is the same whatever the 'entities' in question are.

We will therefore proceed by using the formal apparatus of Price (1972), which does not involve any assumptions about the entities under discussion, the characters they possess, or the way in which they (metaphorically or literally) 'reproduce'. While considering a number of interpretative issues—e.g. to which entities and populations thereof it should be applied—we shall then examine how it relates to the aim of science *qua* activity, and how we should understand scientific progress.

It should be noted, however, that the key philosophical arguments in this chapter are independent of Price's equation, which is used primarily for illustrative purposes. The primary reason for using the equation is that it is an actual scientific tool.

2. THE AIM OF SCIENCE

Before we proceed to look at Price's equation, we should first be clear about what 'the aim of science' means, and how to determine, generally, whether *x* is (or is not) such an aim.

Strictly speaking, to talk of 'the aim of science' is plausibly to make a category error. People have aims, whereas activities do not (except perhaps derivatively), and the aim of any person in undertaking any particular activity need not correspond to its 'aim' in the technical sense under discussion in this chapter. But we inherit the terminology from the existing literature on scientific progress, e.g. van Fraassen (1980, pp. 8–9), Newton-Smith (1981, ch. 3), Popper (1983), Laudan (1984), and Sankey (2000).

In the technical sense, 'the aim of science' means, to a first approximation, what we *should* expect science to achieve. Van Fraassen (1980, pp. 8–9) explains this by asking us to compare the aims of chess with those of chess players. Similarly, consider pinball. The aim of the game is to amass as high a score as possible (when 'possible' is suitably interpreted), although one might play simply because one enjoys the experience, or to distract oneself from troubling thoughts. These personal aims are at best auxiliary to the proper aim, and only count as auxiliary if they *contribute to* fulfilling the proper aim. So that's to say that one of the auxiliary aims of pinball is to enjoy while playing *if and only if* that contributes to amassing as high a score as possible. (Note that this is not to say that any given activity only has one proper aim. The possibility that science has multiple proper aims is left open.)

We might extend the analogy a little further by noting that one's success as a pinball player will ultimately be measured not by how much one enjoys the experience, but instead by the highest score one has amassed (or perhaps by the highest score one can consistently amass). The analogy with science becomes somewhat strained at this point, because such a wide range of activities, across a range of disciplines and sub-disciplines, count as 'doing science'. Perhaps, however, we may restore its force by considering a team game, such as soccer, where players in different positions perform different functions. (One of the functions of a centre forward is to score goals against the opposing team, whereas one of the functions of a goalkeeper is to prevent the opposing team scoring goals, and so forth.) Now if we say that performing those functions contributes to achieving the aim of soccer at the team level—to score more goals than one's opponents in each match, say—then we may suggest that one's success in fulfilling one's assigned function(s) may be used, derivatively, to measure one's success *qua* soccer player. One's fame and fortune are irrelevant in making such a measurement (although these may be what one is really after), except in so far as they are indicators of how well one performs those assigned functions.

We now come to the question of how to evaluate whether some *x* is the proper aim—or one of the proper aims—of a peculiar activity. Crucial, for these purposes, is whether there is an appropriate link between method and proposed aim; as Reichenbach (1938), Newton-Smith (1981), and Laudan (1984) have emphasised. For instance, it would not be reasonable to maintain that the aim of chess is to give checkmate to one's opponent if the rules of the game were changed such that this was impossible to achieve (e.g. because a rule was introduced that allowed the king to 'teleport' to any square when doing so would not place it in check, and capture any piece on that square). In short, if there is no scope to achieve *x* within the boundaries of the game (i.e. by following the rules of the game) then *x* cannot be an aim of the game. Note that this is so even if any given player mistakenly thinks that *x* can be achieved within the boundaries of the game, and plays the game in order to achieve *x*. So even if Laudan (1990, p. 315) is correct that 'rules are best seen . . . as proposed means to the realization of desired ends', a rule which is evaluated as appropriate from a first-person perspective may be inappropriate as a matter of fact.

On the other hand, the mere possibility of achieving some end while performing an activity does not mean that said end is a proper aim of the activity. It is possible to find love while shopping, but it does not follow that one of the aims of shopping is to find love. Similarly, van Fraassen does not deny the mere *possibility* that science might arrive at the truth, the whole truth, and nothing but the truth. But this does not lead him to conclude that the aim of science is truth. The means would have to be *sufficient for*, or at the very least have a good chance of achieving, the end. So if *x* is an aim of science given method *y*, then doing science by *y* must give a high probability of achieving *x*. Let's call this the strong view of aims.[2]

Before we rest content with this understanding, however, we should consider an objection. Think of Olympic athletes, many of whom compete with the aim of winning gold medals. Since the vast majority will be incapable of achieving said end, except under the most exceptional circumstances, should we not say that they are behaving irrationally?

In answering this question, we should first remember the prior distinction between first-person and external perspectives. If a given athlete strongly believes that she can win a gold medal, then we should not say that she is irrational even if she has no chance whatsoever (given any possible training regime); that is, assuming she has reason to believe (falsely) that she can achieve a gold medal. However, if we were to point out to her that her belief was false (and she recognised the force of our argument), then we would expect her to trim her aims accordingly. And in saying this, we need not deny the motivational advantage that might nevertheless be gained by *aspiring after* a gold medal. (Similarly, van Fraassen might agree with Popper that the quest for truth is *motivationally* important, without conceding that truth is the aim of science.) But note that such an aspiration would only be good for the athlete if there were some utility for her in improving her performance yet failing to win gold. If we knew that failure to achieve gold if she continued training would lead her into depression for the remainder of her days, we would instead urge her to give up (competitive) athletics.

But would the point go equally if we could only show the athlete that she had a very small chance, e.g. a probability of 0.05, of winning gold? It would not, we might think, provided that the athlete could *reliably* make progress towards that goal by her training regime, i.e. continually improve her performance (despite occasional downward fluctuations due to injuries, improper diet, and so forth) and work her way up the rankings. We may add that in most actual cases, athletes derive satisfaction from approaching the goal even if they never reach it.

We might therefore conclude that x may be the aim of science even if the probability of science achieving x is low, *provided that doing science is a reliable means by which to make continual progress towards* x.[3] (But x is certainly not the aim of science if it is not possible for science to achieve x, or if doing science provides no reliable way to make continual progress towards x.) Let's call this the weak view of aims.

In what follows, we will work on the assumption that either view of aims, strong or weak, may be correct. This is simply to be as open-minded as possible in our consideration of alternatives.

3. PRICE'S EQUATION

Consider a population P, containing n entities which vary with respect to an attribute z. Let z_i represent the value of the attribute for entity i, and $\bar{z}$ represent the average value of the attribute for the entities in P, such that:

(1) $\bar{z} = \frac{1}{n}\sum_1^n z_i$

Let w_i signify the number of entity i's offspring, or the *absolute fitness* of i, and $\bar{w}$ signify the average number of offspring of the entities in P, such that:

(2) $\bar{w} = \frac{1}{n}\sum_1^n w_i$

Finally, let z_i' denote the average value of the attribute for the offspring of entity i. The *transmission bias* of entity i—which measures how faithfully the attribute is transmitted by i—is:

(3) $\Delta z_i = z_i' - z_i$

And the average transmission bias can therefore be represented as:

(4) $\overline{\Delta z} = \frac{1}{n}\sum_1^n \Delta z_i$

Now consider the population O, which consists of the offspring of the entities in P. The average value of the attribute under consideration in O, $\bar{z}_o$, is captured by the following crucial equation:

$$(5)\quad \bar{z}_o = \frac{1}{n}\sum_{1}^{n}\frac{w_i}{\bar{w}}z_i'$$

As Okasha (2006, p. 21) explains, the average value of the attribute in O is therefore calculated by considering each individual entity in P, multiplying the fraction of entities in O it produces with the average attribute value for that fraction, and summing.

We are now in a position to derive a value for the change in average attribute value from P to O, which is the core quantity that we are interested in, intuitively, from an evolutionary perspective. From (1) and (5), this is:[4]

$$(6)\quad \bar{w}\Delta\bar{z} = \mathrm{Cov}(w_i, z_i) + \mathrm{E}(w_i\Delta z_i)$$

When both sides are divided by , this gives:

$$(7)\quad \Delta\bar{z} = \mathrm{Cov}(\omega_i, z_i) + \frac{\overline{\Delta z}}{\bar{w}} \text{ where } \omega_i = \frac{w_i}{\bar{w}} \text{ (relative fitness)}$$

Prima facie, one might describe $\mathrm{Cov}(\omega_i, z_i)$ as capturing the element of variation due to selection, and $\frac{\overline{\Delta z}}{\bar{w}}$ as capturing the element of variation due to transmission bias. But even when we put to one side the possibility of correlations between ω_i and z_i in circumstances where the value of the attribute has no causal bearing on the fitness, we might find it curious that the element due to transmission bias should depend on fitness. As Okasha (2006, p. 26) points out, however, we may instead consider:

$$(8)\quad \Delta\bar{z} = \mathrm{Cov}(\omega_i, z_i') + \overline{\Delta z}$$

Okasha (2006, pp. 29–31) prefers (8) from a causal (but not statistical) perspective, on the basis of considerations relating to the different ways in which there could be no selection on z. In what follows, we will follow suit.

In plain language, for those not enamoured by mathematics, Price's equation as expressed in (8) involves two main factors: *transmission* and *selection*. $\overline{\Delta z}$ reflects the change (between two populations) that would happen if no selection were present. Consider, for example, the way that an increase in the percentage of brown-eyed humans may occur, across generations, due purely to genetic considerations. $\mathrm{Cov}(\omega_i, z_i)$ represents the additional change due to selection. So if brown-eyed humans were the target of an especially nasty and effective predator that didn't care for other humans, then the overall percentage of brown-eyed humans might decrease in spite of genetic considerations. Note that the two factors may also counteract one another; thus the percentage of brown-eyed humans may remain the same although selection occurs.

4. APPLYING PRICE'S EQUATION: CRITICAL RATIONALISM AND TRUTH AS THE AIM OF SCIENCE

In the context of evolutionary epistemology, theories are the sorts of entities with which authors are typically concerned.[5] But this leaves many questions open. First, what attributes should we be interested in? There are three obvious possibilities, when we consider standard accounts of scientific progress: truth, truth-likeness, and empirical adequacy (whether in degree or not).[6] I will also say something about the slightly less obvious structural adequacy.

Second, how should we differentiate between populations—e.g. understand 'reproduction' and 'offspring'—in such a context? Note that the entities in O need not be of the same type as those in P, provided that they possess the relevant attribute, as noted by Rice (2004). But this leaves the possibility, for example, of letting the O entities be statements logically derived from the theories in P.

A concrete example, which to the best of my knowledge involves a novel evolutionary analogy, might help. Let the parent be classical mechanics—taken, roughly, to consist of Newton's laws of motion and gravitation—and let the offspring be the predictions derived from this theory (or theoretical framework). Let $z = 1$ if the entity under consideration is true, and $z = 0$ if the entity under consideration is false. 'Reproduction' occurs when the parent is used to generate statements, e.g. that the best angle at which to throw a javelin when on a level surface on Earth is forty-five degrees (in order to maximize the horizontal distance it travels), or that Halley's comet will next appear in the sky in such and such a year. As such, the theory has the potential to have an infinite number of 'offspring', and will continue to 'give birth' to new statements for as long as we use it.[7]

Thinking about this example just a little further, however, leads us to realize that matters are really much more complicated. 'Reproduction' requires further statements—auxiliary hypotheses and/or statements of initial conditions, in light, as we have seen in Chapter 5, of Duhem's thesis—and might therefore be better understood as 'sexual' (and not limited to only two partners)! So if we understand a theory to consist simply of a collection of statements—for the time being, let's now do this—we may also expand P to include these auxiliaries and statements of initial conditions.[8] This will limit the number of possible offspring, although the actual number may still be determined by the logical and mathematical operations we perform on the entities in P in the actual world. Note also that if we like, we can consider 'reproduction' to occur only if the relevant operations are properly applied; human errors in calculation may then be understood as failed attempts to reproduce (which nevertheless issue in statements which might be *mistaken* for offspring of the entities in P). In effect, this allows us to put miscalculations to one side.

Now let's consider how $\bar{z}_o$ relates to $\bar{z}$, and what the average transmission bias ($\overline{\Delta z}$) is, in a variety of circumstances. If $\bar{z}$ is unity, then $\bar{z}_o$ is also unity—and $\overline{\Delta z}$ is zero—as the reproductive process ensures that true parents have true children. If $\bar{z}$ is zero, however, then $\bar{z}_o$ may have any value between zero and one (and the same goes for $\overline{\Delta z}$); false parents may have either true or false children. Finally, if $\bar{z}$ is greater than zero but less than one, a similar result holds; equally, children with true and false parents may themselves be either true or false. This is not, perhaps, terribly interesting. But we can say a little more if we let the O population instead be the result of all *possible* (legitimate) operations applied to members of the P population (so that fitness is fixed for any given population). In this event, $\bar{z}_o$ is unity *if and only if* $\bar{z}$ is unity. Else, the value of $\bar{z}_o$ will plausibly be fixed at a different value (since the population is infinite).[9]

We have now shown what anyone who understands elementary deductive logic should already know; that generating valid arguments does not serve to 'select' true statements (offspring), but only to transmit truth if it is present in all the premises (parents). *Prima facie*, this illustration is rather unremarkable. But when we consider a derivative issue, namely whether adopting the approach advocated by Popper (1959) can select false universal statements (so that these can be rejected), it proves much more interesting. Specifically, it throws into sharp relief that unless we think that we have a reliable source of 'basic' (or observation) statements—e.g. a source that produces true perceptual beliefs with a much higher relative frequency than false perceptual beliefs—there is little reason to suppose *even* that our knowledge of which theories are false will generally increase over time (as Watkins [1997] suggests it will).

In order to make this manifest, let the P population contain all our past and present universal hypotheses, and the O population contain all those past universal hypotheses that have been *classified* as false due to their inconsistency with accepted observation statements. Again letting a z-value of 1 represent truth and a z-value of 0 represent falsity, we can now ask whether we should expect $\Delta\bar{z}$ to be negative (on the reasonable assumption that some of our past and present universal hypotheses, e.g. 'All men are mortal', are true). First, note that a past or present universal hypothesis only 'reproduces' if it comes into conflict with—or if preferred, 'mates with'—an accepted basic statement (e.g. as a result of a test). Second, as we have already seen, note that the 'offspring' will be false if the 'partner' basic statement is true, but may be either true or false if said basic statement is false. Now if our basic statements were only ever true, *then true universal hypotheses would never reproduce*; i.e. there would be no average transmission bias, $\overline{\Delta z} = 0$, and $\Delta\bar{z}$ would be negative provided that $1 > \bar{z} > 0$ and O had at least one member. By actively identifying inconsistencies and thereby increasing the number of entities in O, e.g. by testing our theories, we would therefore be making scientific progress; and it would be quite legitimate to say that the aim (or at least one aim) of science is to identify and rule out false theories. (There may be some false theories which cannot be ruled out, however; we shall come back to this.) It is also easy to see that something similar would be the case if a very high proportion of our observation statements were true, but not if a very high proportion were false. The details of in-between cases, and where an appropriate cut-off point lies, need not concern us here. But note that if observations/experiments had a high *propensity* to result in true beliefs capable of falsifying universal hypotheses then this would also suffice; a run of mistaken falsifications would then be a matter of bad luck. Perhaps this is why Popper (1974a, p. 1114) states, at one stage at least, that:

> Our experiences are not only motives for accepting or rejecting an observational statement, but they may even be described as *inconclusive reasons*. They are reasons because of the generally reliable character of our observations; they are inconclusive because of our fallibility.[10]

Even if our (identification of true) observation statements were infallible, however, we would still not be entitled to say that the *aim* of science conducted in a critical rationalist manner—rather than a mere *aspiration* thereof—is to identify true theories. This is because increasing the number of theories correctly identified as false need not reduce the number of theories (provisionally) classified as true which are actually false. There is no reason to expect that replacement theories will be true, if we forswear any appeal to inductive reasoning (or the context of discovery). Furthermore, a false hypothesis successfully discarded could be replaced by many new false hypotheses, and new hypotheses sometimes arise which are not merely replacements for falsified ones. (Consider hypotheses concerning newly discovered sorts of entities.) In short, successful falsifications are insufficient to ensure, or even to raise the probability, that successive generations of theories will increase in $\bar{z}$ value.

We have also made it clear—by implication—what would need to be the case in order for Popper's prescribed methodology to be well suited to finding/identifying true theories. (But note that it may be worthwhile to identify and rule out false theories, in the hope that they might be replaced by true ones, even if this is all that we can reliably do.) In addition to a reliable source of observation statements, we would require something like a reliable source of *replacement* hypotheses, at the very least. (If when we weed out a false theory the replacement is likely to be true, then we have what's needed.) Popper (1959, p. 32) appears to deny precisely that there is such a source, or at least that we have any reason to believe that there is, when he writes: 'there is no such thing as a logical method of having new ideas, or a logical reconstruction of this process.' The point is that even if there is such a source, we have (on Popper's view) no ability to tell what it is. So we can hardly say with any reasonable confidence that science (conducted according to any peculiar method) involves such a source.

In light of this, one might think that allowing for induction is the way to resolve the tension between method and professed aim; that only allowing theories with high inductive probability to enter the scientific corpus would ensure progress toward the truth. But this would only be correct if a conclusion with a high inductive probability (based on true premises) were also to have a high aleatory probability—e.g. propensity to be true in the actual world, or relative frequency of truth across (logically) possible worlds.

As we have already seen in Chapter 2, there does not appear to be any successful argument that this is so. A high degree of rational belief in *p* according to a subjectivist or an Objective Bayesian may not, that is to say, correspond to a high chance of *p*. Imagine all those similar possible worlds in which we roll a particular die and get one hundred sixes in a row. Is the die unfair in most of those possible worlds? Our intuitions are silent, not least of all because there are infinitely many such worlds.

Those who would invoke induction in order to defend the notion that the aim of science is truth—or even, as we will see, empirical adequacy—would have to solve this problem in order to be able to use an evolutionary analogy successfully in support.

5. VERISIMILITUDE TO THE RESCUE?

One might instead suggest that the process of testing theories will tend to result in an increase in verisimilitude (or truth-likeness) of our theories over time; and indeed, Popper (1972, p. 57) at one point suggests 'the search for verisimilitude is a clearer and a more realistic aim than the search for truth'.[11] (This appears to suggest that Popper would have rejected the weak view of aims; because if the weak view were correct then one could defend the thesis that the aim of science is truth by showing that practising science can reliably produce theories of increasing verisimilitude, on a generation by generation basis. That is, given the less controversial assumption that it is possible to arrive at the whole truth.) It is easy to see that we can understand such a suggestion by introducing degrees of truth-likeness, represented by a continuum of *z*-values; so if T_1 has a *z*-value of 0.999 then it is almost true, and also considerably more truthlike than another theory, T_2, with a *z*-value of 0.6. How we could (or should) make such comparisons is, notoriously, a matter of considerable controversy. Miller (1974) and Tichý (1974) independently showed that Popper's account of verisimilitude was fundamentally flawed, and subsequent accounts—see Miller (1994), Niiniluoto (1998), and Psillos (1999) for an overview and discussion—are not terribly convincing either.

But we shall try to avoid this problem by discussing examples where the notion of a truthlike answer makes intuitive sense.

Imagine that our task is to discover the relationship between two positive variables, x and y, and let this be known to be a linear relationship: $y = mx + C$. Imagine also that m is 1 and C is 0 (unbeknownst to us). Intuitively, $y = x + 0.1$ is closer to the truth than $y = x + 0.2$, and $y = 1.1x$ is closer to the truth than $y = 1.3x$. Similarly, as illustrated in Figure 7.1, the case might be made that $y = x + 1$ is considerably closer to the truth than $y = 1.1x$, despite the fact that the latter will be a better approximation to the truth in a particular range, i.e. between 0 and 10 (as illustrated in Figure 7.2).

But now imagine that we only have a number of data points in the region $10 > x > 5$, to which $y = x + 1$ is an excellent fit. (Imagine also, for the sake of argument, that the data has been gathered perfectly; that any errors are due to practical limitations, e.g. with the measuring apparatus, and that the actual values for x and y fall within the error bars.) We therefore adopt the hypothesis that $y = x + 1$, and in order to test it—because we are good Popperians, perhaps!—we begin to investigate the region $5 > x > 0$. After collecting some more data, we discover that $y = x + 1$ is false. Instead, $y = 1.1x$ now provides the best fit to our combined data, provided we continue to work on the assumption that the line must be straight (which just so happens also to be right). How this could happen should be obvious from Figure 7.2. Unbeknownst to us, however, our new hypothesis is less truthlike than our previous one.

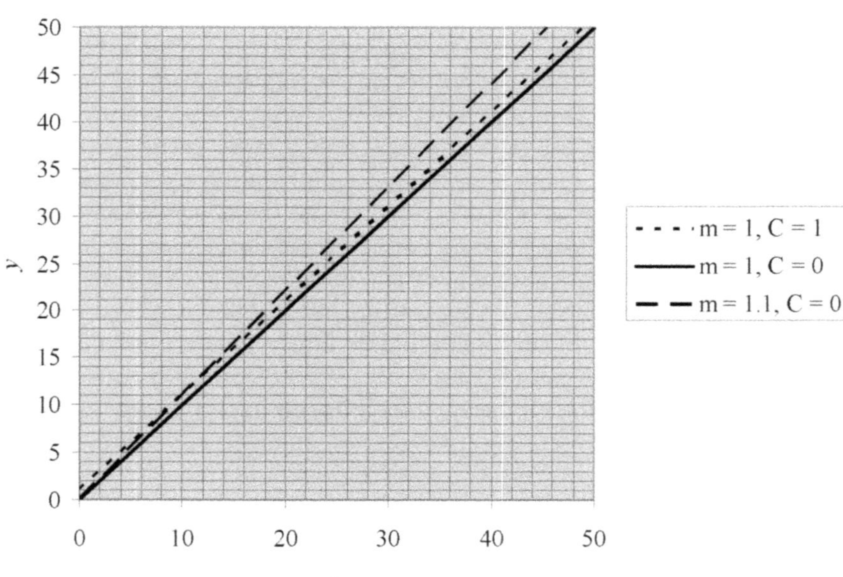

Figure 7.1 Lines over an interval.

What this shows is that it is perfectly possible, by testing a theory, to reject a more verisimilar option in favour of a less verisimilar one; and that this is so even when there is no error on the part of the experimentalist. Note also that this might continually occur, over successive generations of hypotheses. More generally, indeed, an infinite number of curves can connect any finite number of data points, and we cannot always know that we should expect a particular shape, e.g. a hyperbola or a straight line. We could be investigating a region where a relationship appears linear when it is not, and so forth. (Think of how temperature difference and potential difference relate to one another, for a range of thermocouples.) We should therefore concur with Miller (1994, p. 200) that:

> [W]e must not be misled . . . by what is a very natural, but unfortunately not correct, picture of how the verisimilitude of our hypotheses might increase—I mean the idea that we might systematically eliminate errors one by one, thus gradually approaching total freedom from error.

By extrapolation, and with reference back to the earlier discussion of the aim of science, we can see that there is not even any reason to think that there is a reliable means by which to *approach* an error-free situation. The selection process does not favour theories with high verisimilitude over those with low verisimilitude, but only those that fit the finite data. So we have no reason, in general, to expect to move closer to the truth with greater frequency than we move away from it.

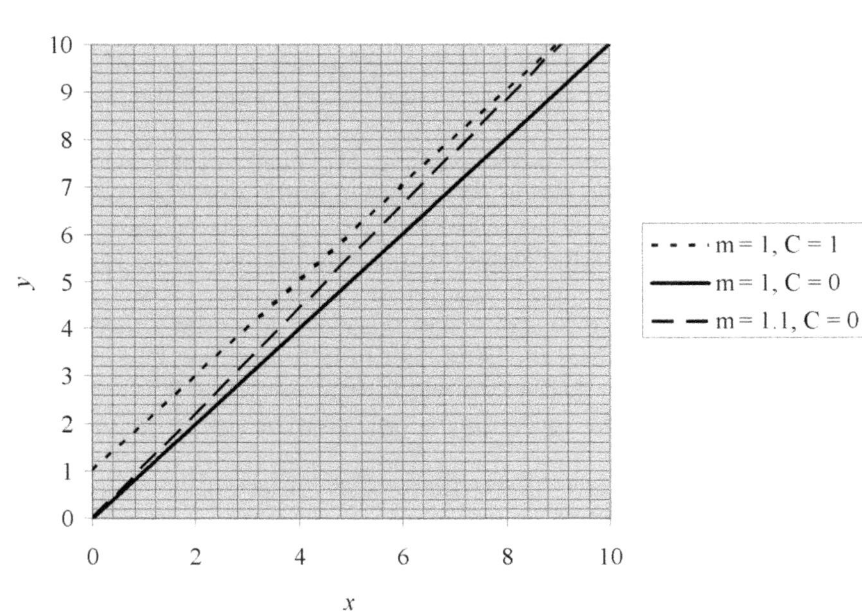

Figure 7.2 Lines over a sub-interval.

We have now seen that even if experience *and* experimental results were infallible—and they are not!—there is no reason to expect the process of testing theories to lead us (reliably) to adopt ever more verisimilar theories. Among other things, the order in which evidence is gathered can affect $\Delta\bar{z}$ in highly unpredictable ways, between successive populations of accepted theories (while not necessarily affecting how many hypotheses we successfully classify as false). In short, verisimilitude does not 'come to the rescue' as an aim of science (on a strong view of aims) or save truth as an aim of science (on a weak view of aims). Provided the evolutionary analogy is apt, it is wrong to say 'We learn from experience by repeatedly positing explanatory hypotheses and refuting them experimentally, thus approximating the truth by stages' (Gattei 2009, p. 43).

6. THE NEXT CONTENDER: EMPIRICAL ADEQUACY

One might nevertheless think that empirical adequacy—where 'a theory is empirically adequate exactly if what it says about the observable things and events in this world is true'[12] (van Fraassen 1980, p. 12)—survives as a defensible aim on the basis of an evolutionary analogy. In order to see whether this is so, we can let $z = 1$ for an empirically adequate theory and $z = 0$ for an empirically inadequate theory. If wished, we can also consider degrees of empirical adequacy, just as we previously did degrees of truth-likeness, with $z = 1$ denoting (complete) empirical adequacy and $z = 0$ denoting *complete* empirical inadequacy. It seems likely, however, that generating a plausible account of degrees of empirical adequacy would prove just as difficult as generating an unproblematic account of degrees of truth-likeness. This is suggested by the fact that empirical adequacy is defined in terms of truth.

Let's consider the curve-fitting example, explained in the previous section, again. Given the limitations of our instruments, more than one curve (and relevant theory) might prove satisfactory—e.g. from the point of view of empirical adequacy with respect to the readings on the measurement apparatus—even for an infinite number of experiments. Thinking of the limitations of measuring devices makes this clear. Consider a theory that successfully predicts where the hand on a fully functioning kitchen clock will be as far as we can discern without instruments. It need not establish the truth about the time (according to our conventions) in the order of milliseconds.

However, when we give full modal force to 'observable'—on a related note, see Ladyman (2000)[13]—it becomes evident that a theory which appears to be empirically adequate today may prove not to be tomorrow, e.g. when a new measuring apparatus is developed.[14] Think, in this regard, of special relativistic predictions concerning time dilation, corroborated by Hafele and Keating (1972) and Allan et al. (1985), which it would not have been possible to test without clocks that can measure differences in nanoseconds. Moreover, we might let x and y be directly observable; be *readings* on a specific class of barometers at heights (on Earth) which can be determined and discussed without any appeal to theoretical entities, for example. And this indicates that the same problems identified previously also apply to the claim that (increased degrees of) empirical adequacy will be achieved by selection. Any empirically adequate theory must get the curve shape exactly right, and while we can be reasonably confident that gathering more evidence will lead us to identify ever more empirically inadequate theories—that is, to repeat, on the condition that our observations are reliable—there is no reason to suppose that this will get us any closer to establishing an empirically adequate alternative. What's more, our replacement theories may turn out to be less empirically adequate than their forerunners (considering all the possible evidence). Being faced with an infinite number of possibly empirically adequate theories is little better than being faced with an infinite number of possibly true ones.

Might one not say, nevertheless, that the aim of science is to rule out empirically inadequate theories? At first sight, it is hard to see the point because although a theory may be empirically adequate without being true, *any empirically inadequate theory must be false.* So if we can succeed in identifying empirically inadequate theories, we can also succeed in identifying false theories. In formal terms, we are interested in two properties, $z(t)$ and $z(e)$, which can be possessed by the same type of entity. If $z(t) = 1$ then the entity is true, if $z(e) = 1$ then the entity is empirically adequate, and so on. For any such entity, if $z(e) < 1$ then $z(t) < 1$.

However, we may also consider entities that have $z(t)$-values without having $z(e)$-values, which are true or false (or verisimilar to a specific degree) without saying anything 'about the observable things and events in this world' (i.e. without being empirically adequate to any degree). So the aim of science, the critical rationalist and constructive empiricist should agree, is *primarily* to rule out those theories that are empirically inadequate.[15] Neither provides any reason to expect that science can rule out false theories that say nothing about the observable (if there are such things).

7. HOW ABOUT STRUCTURAL REALISM (OR STRUCTURAL EMPIRICISM)?

Given the current popularity of structural realism, it also deserves a mention. The basic idea behind this view, as it was initially proposed by Worrall (1989), is that mathematical structure is often carried over in theory change (at least in so far as the equations used in older theories are 'limiting cases' of those used in their successors). It is superior to scientific realism in so far as it does not suggest that science makes ontological progress across generations; and therefore Laudan's (1981) pessimistic meta-induction, based on the idea that science past has posited all sorts of entities that we no longer believe in, such as caloric and phlogiston, is no argument against it.

So might we say that the aim of science is to capture the structure of the world, or even just the proper mathematical structure for describing it? Again, the previous curve-fitting example applies. The correct mathematical equation is clear. The fact that there is no reason to suppose that we will move towards that equation rather than away from it as we investigate the data is equally as clear. To repeat, the selection process favours those theories that fit the finite data. And this holds even if we simply stipulate that any replacement theory must account for most or all of the same data that the previous theory did (e.g. that a 'correspondence principle', as discussed in Chapter 2, must hold). We may, of course, end up with a theory that nicely fits all the data we have through such a process. *But there's no reason to expect it to be structurally adequate any more than empirically adequate.*

Many alternative forms of structural realism, and derivatives, have been spawned—see Ladyman (2009) for an overview—but perhaps the most notable, for current purposes, is the structural empiricism championed by Bueno (1999, 2000) and more recently van Fraassen (2006, 2008). It should already be easy to see that this succumbs to the previous objection too; but to emphasise the point, it is worth pointing out that it is simply a peculiar 'version . . . of constructive empiricism' (Bueno 2008, p. 213).

8. CONCLUSION

Working on the assumption that evolution by selection can be illustrated via Price's equation, we have drawn a number of conclusions about whether scientific realism or constructive empiricism (or even structural realism) can be adequately motivated by appeal to an evolutionary analogy.

First, we have seen that a critical rationalist cannot use an evolutionary analogy to establish that the aim of science is either truth *simpliciter* or the generation of theories of increasing verisimilitude. Second, we have seen that it is only possible to show that the aim of science is to identify false theories, by use of such an analogy, if one accepts that there is a reliable source of (true) observation statements. Third, we have seen that appeal to induction does not provide a simple way to remedy this problem, because inductive probability of *p* (even if it exists) does not appear to correspond to aleatory probability of *p*. Fourth, however, we have seen that empirical adequacy (or even just structural adequacy) does not appear to be a defensible aim either. We concluded that the only aim which can be reasonably established by an evolutionary analogy—on the assumption that we have a reliable source of true observation statements—is the identification of a particular class of false theories (or, given Duhem's thesis, systems of theories), namely those that are empirically inadequate.

Long after reaching this conclusion, I discovered the following passage in *The Poverty of Historicism*:

> The result of tests is the *selection* of hypotheses which have stood up to tests, or the *elimination* of those hypotheses which have not stood up to them, and which are therefore rejected. It is important to realize the consequences of this view. They are these: all tests can be interpreted as attempts to weed out false theories—to find the weak points of a theory in order to reject it if it is falsified by the test. This view is sometimes considered paradoxical; our aim, it is said, is to establish theories, not to eliminate false ones. (Popper 1960, p. 133)

At this juncture, I believe that Popper ought to have said that our aim is, indeed, to eliminate false theories. I think he should have gone on to say that we cannot establish theories, and that this is precisely what his discussion of the problem of induction (and his finding about the logical probability of universal theories relative to any finite data), which we covered in Chapter 2, shows. Rather inexplicably, however, Popper continued with:

> But just because it is our aim to establish theories as well as we can, we must test them as severely as we can . . .

As I argued in Chapter 2, it would be better to say that we test theories as severely as we can to improve our chances of ruling them out *if they are false*. In short, science is like life in so far as it invariably involves making lots of mistakes. The best one can do is to strive not to make the same mistake twice! Subjecting theories to testing is typically a way to err in a 'safe' environment, i.e. one where there are no serious negative consequences if the theory fails (in the same way there might be if the theory were being employed to guide decision making in a life or death scenario). That's the main point of the exercise. We are left with theories that we can't seem to find anything wrong with. But that's no guarantee that using them won't prove to be our undoing.

I should add that it would be wrong to reject the notion that an evolutionary analogy is appropriate for understanding the aim of science, especially under the method proposed by critical rationalists. Conjecture and refutation is clearly a selection process, as stated by Popper in the penultimate quotation. Tests are means by which to weed out theories.

It's also worth reiterating that science may still *aspire after* truth, and indeed that the 'critical *quest* for truth' (Popper 1959, p. 281) may be motivationally crucial. I will say something more about this in the final chapter.

ACKNOWLEDGEMENTS

This chapter is based on Rowbottom (2010).

8 Thoughts and Findings

This chapter, with which I conclude, is rather different in flavour from the others. Having looked at various aspects of Popper's critical rationalism (and derivatives) in considerable depth, I want now to take the opportunity to stand back and articulate my view on where it stands (and where I stand) in light of the previous analysis, criticism, and development. For the most part, I will avoid lengthy philosophical analysis—the relevant arguments have already been presented in the previous chapters—but will instead adopt a rather more conversational (and even confessional!) approach. In short, my intent is to say what I think, as bluntly as I can, about what I take myself to have shown.

Allow me to begin by stating plainly that I believe there to be deep inconsistencies in Popper's philosophy which need to be explicitly resolved, one way or another, before it can become an appealing prospect. Key among these is the tension between anti-inductivism (and the associated critical deductive method) and scientific realism. As I will explain in the following, I do not believe that one can have it both ways; and as a result, I personally reject realism (although I do not retreat as far as instrumentalism, which was Popper's primary anti-realist target).[1] If we forswear ampliative inferences—or at the very least, hold that they are not truth-conducive—then our science must aim at something other than truth.

I should also voice the opinion that Popper occasionally vacillated on key issues—e.g. what we should understand the aim of science to be (although he was always of a broadly realist bent) and why we should use observation statements as basic statements—and often neglected, alas, to explain exactly how and why his views changed over his long and distinguished career. For me, these were failings on his part. There were others, on Agassi's (2008) account, which make it clear that Popper did not manage to live up to his own high standards concerning individual rationality![2] (But who could? Think back to the earlier discussion of 'How to be a Pancritical Rationalist' in Chapter 1.)

However, it would be a terrible mistake—and one which has been made in some parts of the Anglo-American philosophical community, for reasons I allude to in the preface, I fear—to overlook Popper's considerable contributions to our objective knowledge (as he would have called it) and the enduring significance of his emphasis on criticism and critical procedures.[3]

(I need hardly add, I hope, that it would be a similar mistake to overlook the work of Popper's ex-students, or philosophers broadly sympathetic to critical rationalism, on related issues. I have had the pleasure of referring to this work throughout.) I hope to have already shown this. In particular, I hope to have shown how and why critical procedures are at the heart of good individual and group inquiry, with my work on corroboration theory, intersubjective probability, and the distribution of functions with respect to collective rationality. I will say a little more about these issues, among others, in the following.

1. SCIENTIFIC REALISM AND ANTI-INDUCTIVISM

If we think that truth is the aim of science, then we must also think that doing science (perhaps in a particular way) can reliably lead us to, or at least towards, the end of truth. But why should we think this, if we proceed in science just by guessing at theories (when necessary) and testing them against experience? As I argued in Chapter 7, doing this would at best enable us to reliably rule out false theories. It isn't even as clear as we would like that we can succeed in doing this; after all, we have to rely on our observation statements being true more often than not!

If one is a realist, then, the most obvious thing to do is to reject the view that we simply guess at theories. If we instead had some sort of mechanism that helped us to propose true (or highly truthlike) theories for testing much more often than false (or highly false-like) theories, then we would be onto a winner. Every time we rejected a false theory, we could be confident that we'd replace it with a true, or at least highly truthlike, theory. (That is, if a replacement was required.) We would be on an inexorable march towards truth, the whole truth, and nothing but the truth. We would be sure to get closer to our target destination, and to reap the benefits for so doing, even if we never completed the long journey. (We could think of it as a journey from the desert into fertile land, with the terrain continuously improving all the way.)

As I argued in Chapter 2, it is dubious that appeal to ampliative inferences can help in this task (because high inductive probability does not correspond to high frequency of truth). But it is clear that deduction alone—and indeed criticism more generally—will not enable us to spot true (or good), rather than false (or defective), theories. I think this largely explains why Popper's anti-inductivism is so unappealing to most philosophers of science. Being of a realist inclination, they seek some way to show how the method of science can be sufficient for achieving what they take to be its aim. So perhaps it is no surprise that van Fraassen rejects induction, having already rejected realism (although I maintain, as I argued in Chapter 7, that criticism alone will not reliably lead us to empirical adequacy any more than truth).

I am not convinced that this link between anti-inductivism and anti-realism has been sufficiently well emphasised before, even if some philosophers have recognised it, and I take it to be an important finding. *In general, I hold that anti-inductivism and scientific realism—where the latter is understood as a thesis about the aim of science, and not merely the claim that modern science just so happens (perhaps luckily) to be (approximately) true—are incompatible.* I think that critical rationalists should face up to this, and be honest about it. But I also think that more anti-realists should carefully consider whether, and if so why, they need to appeal to induction at all. If it isn't supposed to be propping up realism (or something similar, such as the quest for complete empirical adequacy), then what work is it doing?

I would also remind critical rationalists that the quest for the truth may still be important from a motivational perspective, and indeed that there is no harm in aspiring after truth (because it is possible, at least, for scientific theories to be true).[4] There is a world of difference, however, between hoping we'll be lucky and claiming that some activity has a well-defined aim that there is a strong reason to think we'll achieve by doing it.

2. INDIVIDUAL VERSUS GROUP RATIONALITY

It is perhaps natural, in the contemporary Western world at least, to focus on the individual when it comes to questions concerning rationality. But a move to thinking about these matters at the group level has been under way for a while now, and is illustrated nicely by the existence of (relatively young) journals such as *Social Epistemology*. I believe this is a trend which critical rationalism should embrace, as suggested by the discussion in Chapter 6.

Unfortunately for critical rationalists, one might think, imagination and criticism do not generally appear to be enough to ensure the best possible inquiry when we come to deal with groups. What's more, dogmatic behaviour even appears to be desirable, for the group as a whole, in some circumstances. Sometimes it takes a dogmatist to push a failing theory to its very limits, and teach us something new, because a more critical individual would simply have abandoned it for a promising alternative. Similarly, battles between dogmatists intent on defending their respective theories can generate a good deal of light as well as the inevitable heat. So even if one denies that dogmatic behaviour is ever necessary in inquiry, it can nevertheless achieve many (even if not all) of the same results as a 'cool and detached' critical outlook *at the group level*.

Nevertheless, all is not lost for critical rationalism. For it is the *necessity* of a critical function, and not its *sufficiency*—for group rationality in inquiry, or simply for the best possible science (in almost any conceivable circumstance)—that the critical rationalist may emphasise. As we saw in

Chapter 6, much of the activity performed by dogmatists *may be* critical even if it is limited in scope; sometimes the best form of defence for one's own pet theory, for example, is attack of the available alternatives. And more importantly, someone needs to evaluate arguments between dogmatists in order to determine which theory emerges victorious. Clearly, it would be undesirable to have this done by individuals that were themselves extremely partisan.

I therefore think that the future of critical rationalism lies in analysing and articulating the role that criticism can and should play in science (and inquiry more generally) construed as a group endeavour. And on a related note, the issue of the division of critical intellectual labour—of how critical tasks should be distributed—is especially interesting. These matters are context sensitive to some extent, but I expect that some useful general results could be derived with the use of appropriate thought experiments or historical studies. (And we shouldn't forget that all the criticism in the world is no good without anything interesting to criticise, so the question of how to strike the balance between creativity and criticism—e.g. between theory-construction and theory-testing—is also central.)

Leading into the next section, it would be remiss of me if I did not remind the reader that Popper (1983, p. 87) himself emphasised the importance of intersubjective exchanges: 'We move, from the very start, in the field of intersubjectivity, of the give-and-take of proposals and of rational criticism.' Shearmur (1996, p. 83) even suggests: 'Reason, on which Popper places so much stress, is on his account to be understood as an inter-subjective process, which acts as a retrospective check on what we produce.'

3. INDIVIDUAL VERSUS GROUP PROBABILITIES

On a related note, still broadly on the issue of social epistemology, critical rationalists should recognise the significance of group probabilities—and not just logical (even if they exist) or personal probabilities—in inquiry, especially when it comes to the issue of confirmation (or, more properly, corroboration).

As argued in Chapter 3, group decision making can prove superior to individual decision making (given a suitable *ceteris paribus* clause) on several grounds. For one thing, error correction may be improved. Think of proofreading. Using two proofreaders of equal ability is preferable to using just one of those proofreaders, all else being equal, because each may spot errors that the other misses. Analogously, in science, false information used as evidence—for corroboration, say—may be minimized by relying on group decisions. Imagine that 1 per cent of scientists suffer from deuteranopia but that the rest have normal vision. And let's assume that anyone with deuteranopia will misclassify a particular green object as red. The probability of having the object misclassified as red when selecting a scientist at random, and relying on his observations alone, is 0.01. But relying on the majority decision of any more than 2 per cent of the group will render the classification error free in this respect!

Using intersubjective probabilities is also preferable because this increases available background knowledge. (Ideally, we want *b* in a corroboration or confirmation function to reflect scientific knowledge as a whole; and recall that every single term in such functions is a probability which is conditional on *b*. But if we limit ourselves to using personal probabilities, it will never get anywhere near.) Hardly any individual scientist, nowadays, can be expected to have mastered all of the theories in his own discipline; and he may even have little chance of being cognisant of all of the recent published work in his own area of interest! But when scientists get together, they can share information. As part of this process, crucially, they may offer summaries of how their own specialist knowledge is relevant to some wider problem *even when that specialist knowledge cannot be shared with other members of the group due to constraints in expertise.*

Naturally, the emphasis on intersubjective probabilities (of appropriate groups) fits nicely with the focus on group rationality discussed in the previous section. A key question that emerges is: which intersubjective probabilities should we be interested in, e.g. when it comes to theory evaluation? Should we exclude those dogmatically committed to any of the theories under evaluation from the evaluation group? Can we perhaps involve them with the evaluation group in such a way as to ensure that we benefit from their specialist knowledge (when it comes to *b*) and ability to correct errors (e.g. concerning competitor theories) without also allowing their bias to affect the final evaluation aversely? These questions are 'where the action is' in the contemporary critical rationalist programme as I envisage it.

4. INTERNALISM AND EXTERNALISM

A key aspect of work in the critical rationalist tradition—and this is true of Popper's work more generally—is that it often appears to presuppose internalism.[5] Justification *in an internal sense* is thought to be ultimately unattainable, and this leads to an emphasis on criticism and 'critical rationality' instead. Justification *in an external sense* is never discussed (at least to the best of my knowledge), although naturally something similar is touched upon from time to time (e.g. in discussions of the reliability of observation statements). So I think it is correct to see critical rationalism as a peculiar kind of response to the failures of the *internalist* justificationist programme.

But why do critical rationalists not turn to externalism? Perhaps this is because the externalist programme appears to be altogether too descriptive to address the kinds of problems with which they are concerned. Critical rationalists are worried about our individual and group dilemmas, when it comes to finding out the truth (or at least, following what I argue in the previous chapter, some of the false). So even if they accept that there is such a thing as external justification, they will want to know *how we can identify the means by which to achieve it*. The mindset of the critical rationalist still has an internalist flavour—the critical rationalist is concerned with what (critical) reasons are *accessible* to him or the group of inquirers that he is a part of—even though internal justification is rejected. The critical rationalist is worried about *how* he should inquire, and what accessible reasons he can (and does) have to prefer information from one source over another (say). So even if he accepts that reliable sources provide justification, he still persists in asking how, exactly, he can work out which sources are reliable (or at least which sources are not reliable).

In short, saying that what differentiates a knower of *p* from a mere believer in *p* is the reliability of the means by which the former came to believe *p* is fine as far as it goes, but it does nothing to help us to inquire more effectively (beyond informing us that we should seek such a reliable means). It does not, in short, tell us which means are reliable although we should dearly like to employ such means, e.g. to arrive at basic statements with which to criticise our theories.

Perhaps a brief word of warning is in order, however, because some of the arguments employed by Popper and critical rationalists—especially those against induction and for the zero logical probability of any universal hypothesis relative to any finite evidence—are made from external perspectives. The objection to induction, for instance, is that there is no reason to think it is a reliable (in so far as truth-conducive) strategy for hypothesis formation. (At best, it is a heuristic. It is not even clear that it is a good one.) Thus it is considered to be a valid criticism of a belief-forming procedure that it is unreliable.

5. TRADITION AND THE EMPIRICAL BASIS: A PRAGMATIC TURN

We have to start somewhere, and where we start may clearly have an effect, if our observations are theory laden, on where we end up. Inquiry cannot proceed without premises any more than we can question all our beliefs, or indeed what we normally take ourselves to know, simultaneously. If you were not confident that no disaster is about to befall you, then you would not be reading this passage (unless, perhaps, you thought that reading it would somehow avert the disaster).

So where should we start? With respect to theories, Popper suggests that we look to tradition. (To some extent, of course, this will be unavoidable.) With respect to the empirical basis, Popper suggests—or at least, as we saw in Chapter 2, may be interpreted as suggesting—that we may accept observation statements as true unless there are critical reasons against so doing. The critical rationalist will reject, however, the notion that there is any internal justification (or that there are any evidential reasons) for so doing.[6] Consider Russell's (1912, p. 19) claim that:

> [T]he certainty of our knowledge of our own experiences does not have to be limited in any way to allow for exceptional cases. Here, therefore, we have, for what it is worth, a solid basis from which to begin our pursuit of knowledge.

The critical rationalist is not liable to be convinced. From the fact that we cannot doubt that *p*, it does not follow that *p*. Similarly, from the fact that we cannot conceive of a situation in which not-*p*, it does not follow that *p*. There is not even any argument that the probability of *p* is raised if we cannot doubt that *p* or cannot conceive of a situation in which not-*p*. In short, Russell's 'solid basis' is not so solid at all unless we introduce further metaphysical assumptions which he does not present. And to introduce such assumptions, which are not themselves part of the 'solid basis', will only raise the question of whether the basis is really all that solid.

If we adopt a pragmatic line, however, then there is no worry here. We may recommend classification of hypotheses (or observation statements) as true if we can't find anything obviously wrong with them (or can't conceive of how they could be wrong) on the basis that it allows us to 'get going' in inquiry. (There is an interesting parallel between the treatment of grand hypotheses and observation statements here, which fits with the view that they are all hypothetical in nature.) But we may do so with open eyes, that we may be mistaken even from the start; that by inquiring, we are taking a risk.[7]

Let me be frank. I understand how this might seem unsatisfactory to those who want security in their beliefs, or some significant subset thereof (e.g. beliefs based on perception). I understand that desire. Unfortunately critical rationalism is based on the recognition that we cannot have such security; or rather, that even if we have the security (in the sense of a reliable belief-forming process or processes) we can't necessarily recognise and/or have good accessible reasons for thinking that we have it. In short, as we saw in the previous section, critical rationalism is about trying to make the best of a bad epistemic lot *in an internal sense*.

It is worth adding that although Popper emphasises the significance of tradition and observation, there are important senses in which he thinks that we can go beyond them. So we are certainly not *trapped* in some sort of theoretical framework.[8] In fact, he suggests that our faculty of imagination is limited neither by the conceptual resources we have gained via tradition, nor by the conceptual resources we have gained via experience. Rather, it transcends those boundaries: 'every discovery contains "an irrational element", or "a creative intuition", in Bergson's sense.' (Popper 1959, p. 32)

Unfortunately, however, Popper does not offer any detailed account of how he takes concept formation to occur, and how, in particular, we could generate concepts that go beyond (as Russell would have put it) what we are acquainted with.[9] Naturally if we have seen red things but only ever non-red horses, we can conceive of red horses; indeed the British Empiricists recognised as much. It remains somewhat mysterious, though, how we might understand unobservable entities in any other way than by analogy with observable things or properties. Having introduced a pragmatic element, moreover, it is perhaps dubious that we should care about doing any more than saving the phenomena in an economical fashion. So again, I think that anti-realism (of one form or another) beckons. *Finding the truth requires that we can conceive of the truth, whereas finding (some of) the false only requires that we can put the limited theories that we can conceive of to the test.*

6. FALLIBILISM

This brings me on to my final comment. Fallibilism is a significant element—perhaps even the core element—of the philosophical *zeitgeist*. Philosophers of science and epistemologists usually take it for granted. And by admitting the fallible nature of their claims, and pointing out that other forms of inquiry are no different, contemporary metaphysicians such as Lowe (1998, In Press) now seek to re-establish the legitimacy of revisionary metaphysics.

But it is worth reminding ourselves that Popper and the critical rationalist movement—understanding this broadly, I would include Feyerabend and Lakatos—were responsible for emphasising the limitations of science, and our inquiry more generally, in a time when these were not so widely recognised. As Popper (1983, pp. 259–260) put it:

> Science has no authority. It is not the magical product of the given, the evidence, the observations. It is not a gospel of truth. It is the result of our endeavours and mistakes. It is you and I who makes science as well as we can. It is you and I who are responsible for it.

This was a powerful tonic for any remnants of the excessive epistemic optimism of the Enlightenment period. Even our simplest and most successful theories and our most elementary observation statements may be wrong. And although Popper was certainly not the first to suggest this—take Peirce, given our recent mention of pragmatism, as a case in point[10]—it is nonetheless plausible that he contributed to the current widespread acceptance, in the humanities and social sciences (at least), of this doctrine.

Perhaps there is even a clue about why Popper was a realist, here, and the kernel of an argument against some forms of anti-realism. If our so-called 'basic statements' (concerning observables) are just as hypothetical in character as our statements concerning unobservable things, both being theory-impregnated, then why doubt the latter but not the former? As Churchland (1985, p. 36) argues:

> Since our observational concepts are just as theory-laden as any others, and since the integrity of those concepts is just as contingent on the integrity of the theories that embed them, our observational ontology is rendered *exactly as dubious* as our nonobservational ontology.

However, note that the form of anti-realism that I have advocated in Chapter 7—specifically that the aim of science is to rule out empirically inadequate theories—is impervious to this objection. Indeed, we may rule out particular hypotheses about unobservable things—just as we may rule out particular hypotheses about observable things—in virtue of their observable consequences. On the basis of Count Rumford's observations of cannon boring, and similar experiments, we have ruled out the hypothesis that heat is a substance. On the basis of the Michelson-Morley experiment, and similar experiments, we have ruled out the hypothesis that there are unobservable things called electromagnetic waves *which propagate through a luminiferous aether*. That is, subject to the reservations concerning Duhem's thesis. (In the latter case, for example, one may posit aether drag, or claim that the measurement apparatus is affected by movement through the aether, in an attempt to save the aether hypothesis.)

Notes

NOTES TO CHAPTER 1

1. John Preston has suggested to me that Wittgenstein is not appealing to faith in this passage, but instead intimating that beliefs are ultimately grounded in practices. The question then arises, however, as to whether one should critically examine one's practices with a view to changing them (e.g. if they transpire to be inappropriate given one's ends) *in the absence of justification for those practices*. Note that even if one cannot change some of one's practices directly, one may be able to introduce new practices to counteract their negative effects. Even if one cannot prevent one's mood swings, e.g. because of bipolar disorder, one can nevertheless take medication to reduce their frequency. (And, as we will see, rationality is plausibly a matter of practice, in part if not in whole, for Popper.)
2. Presumably, van Fraassen should have added that an empirical refutation of X would suffice. The important point for present purposes, however, is just that if E+ is an empiricist dogma then no empiricist can believe anything contrary to E+.
3. See *Philosophical Studies* 121; Monton (2007); and Rowbottom and Bueno (In Press B).
4. Somewhat surprisingly, van Fraassen cites Feyerabend, but not Popper, in his discussion.
5. I take it that Descartes's (alleged) suspension of belief in God was like this, if indeed Descartes did succeed in suspending belief in God. (I take it that it is possible to question a belief without suspending the belief. In any event, Descartes chose only to reflect on the belief, i.e. to enter into thought concerning its content; he could not have guaranteed that the process would lead to suspension.)
6. See also, for example, Popper (1940, p. 404):

    > [I]t is the most characteristic feature of the scientific method that scientists will do everything they can in order to criticize and test the theory in question. Criticizing and testing go hand in hand: the theory is criticized from very many different standpoints in order to bring out those points which may be vulnerable . . .
7. In the words of Bartley (1968, p. 43), 'The importance lent to the falsifiability criterion and the demarcation problem by Popper and others distorts his thought.'
8. This raises the issue of Duhem's thesis, that a hypothesis cannot be tested in isolation, which I discuss in detail in Chapter 5. For the moment, suffice it to say that we may need to consider whether *systems* of claims are empirically falsifiable.

9. See Artigas (1999) for further discussion of how Popper related his view to that of Bartley. Artigas (1999, p. 99) suggests that 'the moral decision on which Popper's social theory relies is not an irrational faith . . . It is rational.' Popper appears to have thought that he was not advocating 'fideism' in Bartley's sense of the word. See also the unpublished letter from Popper to Graves in Rowbottom (2004).
10. Albert (1985, p. 163) similarly intimates, in his discussion of theologians, that critical rationalists should avoid being '*critical but nonetheless dogmatic*, critical in the things that are not so important to them, dogmatic in those that seem to be more so.'
11. Perhaps Newton-Smith meant 'confidence', but the turn of phrase was ill-advised.
12. See Kuhn (1996 [1962], pp. 187–191).
13. Popper (2003 [1945], p. 255) also suggested that becoming a critical rationalist may not involve adopting a belief, in so far as he mentioned that the adoption of some behaviour would do. As such, he may have foreseen the stance route.
14. This would fit with the view of van Fraassen (2004a, p. 129) that '"[W]hat is rational is whatever is rationally permitted": *rationality is bridled irrationality*.' See the discussion in Chapter 2 and Bartley (1984, p. 116).
15. Another strategy would be to deny that we have any real choice over the stance we adopt. See Rowbottom and Bueno (In Press A, section 2) for an argument against this.
16. Van Fraassen does not suggest that a stance holder must be committed to maintaining the stance come what may. But he does suggest that the necessary commitment is extremely strong, i.e. such that it need be given up only in exceptional circumstances (e.g. in the light of despair).
17. See, for instance, Cohen (2002, 2005).
18. For more on this issue, and in particular for a possible solution that involves the use of 'hinge propositions' that just so happen to be shared between participants in debate, see Pritchard (2009, Forthcoming). This would be quite similar, in critical rationalist terms, to attempting to use shared beliefs as 'basic'—in the short term—in order to break the deadlock. While this strategy might work in a large number of cases of disagreement as a matter of fact, it will not work for all possible cases.
19. A concrete example of this sort of disagreement occurred between Keynes (1921), who thought that we could intuit the truth of particular probability relations, and Ramsey (1926), who denied this.
20. An interesting alternative formulation is given by Stoneham (2007). See Jäger (2007) and Rowbottom (2007c) for criticisms of this formulation, and Shackel (2008) for a response.
21. On Williamson's view of knowledge, for instance, knowing that *p* is compatible with being unaware that one knows that *p*. In fact, Williamson (2002, p. 94) argues more generally for a thesis of anti-luminosity according to which we are 'cognitively homeless'.
22. On related matters, see Cohen (2002) on so-called 'easy knowledge' (which is certainly a problem for *some* forms of reliabilism). As a flavour, consider the following:

    > Suppose I have reliable color vision. Then I can come to know, e.g., that the table is red, even though I do not know that my color vision is reliable. But then I can note that my belief that the table is red was produced by my color vision. Combining this knowledge with my knowledge that the table is red, I can infer that in this instance, my color vision worked correctly. By repeating this process enough times, I would seem to be able to amass considerable evidence that my color vision is reliable, enough for

me to come to know my color vision is reliable. Vogel calls this process "bootstrapping". Clearly I cannot use bootstrapping to acquire knowledge that my color vision is reliable. But the reliabilist appears unable to explain why I can not. (Ibid., p. 313)

23. As we will later see, the core externalist insight may be adopted in order to defend the view that we should be critical rationalists, as may a virtue epistemological framework. The present comments should be understood in this context. Note, however, that *on a critical rationalist view the virtues involved would be linked to avoidance of error rather than acquisition of truth.*
24. Admittedly, it may still be difficult to choose between two internally consistent positions which preclude one another, each of which is preferable to the other on its own terms. However, a pancritical rationalist is limited neither to performing immanent criticism nor to launching transcendent criticism while presuming that what he presently believes is true. As such, she is in the best possible position to choose (*ceteris paribus*).
25. See Miller (1994, §4.3) for discussion of several objections that I do not cover; Rowbottom (2009b) tackles Hauplti (1991), which Miller does not discuss.
26. Agassi et al. (1971, app. A) provide examples of 'various techniques for arguing without risking refutation or defeat in an argument'.
27. Moreover, Agassi et al. (1971) suggest that 'essential to Watkins' discussion appears to be an idea quite alien to Bartley's view, the idea, namely, that if a theory emerges from criticism unharmed, then it emerges victorious.'
28. See also the discussion of Miller (1994, p. 89) in the next subsection.
29. On the criticisability of logic, see Bartley (1984, app. 5), Miller (1994, 4.3e), and Nilsson (2006). Since I agree with almost everything that Nilsson (2006) says in his concise paper, I do not discuss the criticisability of logic, which I see as a specific instance of a 'sources of criticism' problem, in detail here.
30. Recall Bartley's (1984, p. 223) statement that: 'I mean that the criticizers—the statements in terms of which criticism is conducted—are themselves open to review.' In fact, a sympathetic reading of this would lead one to conclude that Helm's criticism was anticipated.
31. David Miller has objected that "Why advocate pancritical rationalism?" is not the sort of question that a pancritical rationalist ought to ask, because it is (or looks like) a request for justification. Rather, the appropriate question is "What's wrong with critical rationalism?" But a dogmatist may answer simply with "It is not endorsed by the authority". Ultimately, the verbal formulation therefore matters little. What matters is how we tackle the question. (It is true that "It is not endorsed by the authority" is not a criticism of the *content* of critical rationalism. This suggests that "What are the negative consequences of critical rationalism?" would be a better formulation. But then I fail to see why one should not also ask "What are the positive consequences of critical rationalism?" This is not a request for justification, but for the pros of adopting a stance/policy. Ultimately, we should be interested in the pros and the cons and whether the former outweigh the latter.)
32. The dogmatist is one form of irrationalist. Other irrationalists might be non-critical and non-dogmatic, but simply float freely on the seas of belief. We will not consider these here.
33. According to Popper's view on learning, outlined in *Realism and the Aim of Science*, it is perfectly possible to learn false things (or to do things in bad or inappropriate ways). See Swann (1999).
34. See also Rowbottom (2008d) for a related discussion and a list of further pertinent references.

35. Watkins does not suggest, of course, that we cannot (or that science should or does not) aspire after truth. His point, rather, is that an aspiration is different from a rational aim.
36. Ward Jones has suggested to me that the way 'justificationism' is used here is somewhat different from earlier, i.e. 'having a positive story about one's belief'. As we have seen, however, appeal to a final authority (e.g. self-evidence or experience) appears to be the only way to have such a positive story *if one wants to avoid regress or circularity.*
37. See Rowbottom and Bueno (2009) for a presentation involving truth tracking, following Nozick (1981).
38. Matters will become more complex if other intellectual virtues are allowed for. But one way to understand critical rationalism (or at least a naive form thereof), from a virtue epistemological perspective, is as the view that the only intellectual virtue is being critical. I believe this is a good way to promote discussion between critical rationalists and mainstream epistemologists; talk of epistemic aims and virtues can replace squabbles about justification in some traditional sense of the notion.
39. Sometimes one is said to entertain a proposition if one has any propositional attitude involving it. Clearly, however, this is not the sense of 'entertain' intended by Bartley; for one could just dogmatically hold the attitude, for instance, that the proposition was false.
40. And if I do not stop with these positions, then why not go on to consider whether the marketer means 'cucumbers from Exeter' by 'double glazing'? Thanks for Peter Baumann for this example.
41. Settle et al. (1974, p. 88) agree:

    > The reasons a man may have for refusing to pay attention to an attack upon his views may be quite plausible and yet have nothing whatever to do with whether he would judge that attack to be decisive against his view had he stopped to hear it, although, of course, a dogmatist of a kind . . . may refuse to pay attention to criticism which he fears may be decisive. Even so, there may be conditions under which a decision to refuse to hear a reasonable criticism is reasonable: a dying believer should perhaps be left to die peacefully in his faith.
42. As we will see later, in Chapter 4, the principle of total evidence still holds for a corroboration-based account of scientific method. For example, we want to employ as much of our evidence as possible because of its falsifying potential.
43. I do not discuss such arguments in detail because it is far less plausible that the consequences of taking up free reflection time will be epistemically beneficial than it is that taking up free reflection time is virtuous. In short, to accept epistemic consequentialism is to make the case more difficult.
44. This occurs even as early as in the final paragraph of Popper (1959).
45. Note that sometimes a position may become problematic as a result of other closely related, or derivative, positions becoming problematic. This is crucial if all positions, even foundational ones such as ethical values, are to be included.
46. Perhaps a little more about the exact meaning of 'entertain' is necessary. I take it to involve serious contemplation, involving the admission of *epistemic* possibility, in the sense it is used by Bartley. So I maintain that I could argue that 'Grass is red and the sky is blue' follows from 'Grass is red' and 'the sky is blue' without entertaining the notion that grass is red. *Even if one were to show that I must consider some logically possible world in which grass is red, it does not follow that I admit the* epistemic *possibility that grass is red in the actual world.*

47. This may not be so on an account of 'in-between believing' such as that provided by Schwitzgebel (2001); see Rowbottom (2007a), however, for an argument that this is not necessary. See also Schwitzgebel (Forthcoming) for a response.
48. As Bartley (1984, p. 121) puts it: 'We can assume or be convinced of the truth of something without being committed to its truth'. Settle et al. (1974, p. 86) add:

    Holding a view open to criticism does not imply doubting it (nor vice versa), just as being certain of a statement does not imply that it cannot be held open to criticism, though it may imply that we do not expect criticism to succeed. Thus we may seriously doubt a theory which we find hard to criticize; or we may be convinced of a theory, which is continually under attack.

49. Perhaps Bartley should have used only 'addicted', rather than 'committed' and 'attached'; many would take it that there is a link between degree of belief in a proposition and degree of commitment to its truth. Here we are more concerned with *commitment to truth come what may*. This is consonant with the proposed removal of (4), discussed in the following; but alternatively, if talk of commitment (as opposed to irrational commitment) is to be avoided, (4) may be jettisoned and (3*) may be retained.
50. See Gillies (1991) and Chapter 3.

NOTES TO CHAPTER 2

1. I should add that Popper (1959, p. 16) did not believe in the existence of a scientific method above and beyond 'the one method of all *rational discussion*, and therefore of the natural sciences as well as of philosophy. The method I have in mind is that of stating one's problem clearly and of examining its various proposed solutions *critically* . . . '
2. When we are faced with a choice of several theories 'on the table' at the same time, the aforementioned criteria might be employed to rank them.
3. Popper (1983, Part I, §3) also believed that we do not, as a matter of fact, learn by induction (and particularly by repetition). However, he added (1983, pp. 38–39): '[E]ven if all of us who deny the existence of "inductive procedures" are wrong, it would be the height of dogmatism to assert that these disputed "facts" create standards of reasoning whose validity is not open to further discussion.'
4. I agree with Popper's reading of Bacon, according to which: 'Bacon described . . . unconscious assumptions as idols and as prejudices' (1983, p. 14). See also Popper (2002 [1963], pp. 17–20). For a different view, see Urbach's discussion (especially p. xix) in the introduction to Bacon (1994 [1620]).
5. See also Popper (1981, p. 88): '*there is no such thing as instruction from without the structure*, or the passive reception of a flow of information which impresses itself on our sense organs.'
6. See Ellis and Lierse (1994), Mumford (1998, pp. 216–217, 236–237), Bird (2001), Mumford (2004, pp. 103–104), and Hendry and Rowbottom (2009).
7. See, for example, Cohen et al. (2003).
8. See Psillos (1999, p. 88) for a reconstruction of Carnap's argument.
9. Moreover, inductivists such as Salmon (1965, p. 268) argue that we should trust deductive inferences because we do not have any reason to doubt that they are truth preserving. The same cannot be said, of course, of inductive inferences. I will argue later, furthermore, that we have reason to doubt that they are even reliable.

10. Besides, advocates of ampliative inference are typically not deductive sceptics. In fact, many propose that deduction is a special case of induction. Consider, for example, the following passage from a call for papers for the PROGIC 2005 (combining probability and logic) workshop:

    > [P]robability generalises deductive logic: deductive logic tells us which conclusions are certain, given a set of premises, while probability tells us the extent to which one should believe a conclusion, given the premises (certain conclusions being awarded full degree of belief).

    The deductivist rejects this view, while maintaining that deductive inferences are, nonetheless, truth preserving.
11. Chalmers (Forthcoming, ch. 1) suggests that there are counterexamples to almost any such definitional claim, and in this case that a consideration of homosexual men with long-term partners suffices to render the statement dubious. (And so, for that matter, might the case of a married man who is separated from his partner and lives alone.) Even if we accept the force of the counterexamples, however, "All bachelors are men" will do nicely for my purposes. In short, my point does not depend on our being able to define 'bachelor'.
12. On heuristics, see Post (1971), Radder (1991), and French and Kamminga (1993).
13. One may doubt that the rigid demand for 'full explanation' is reasonable especially if the successor doesn't entirely overlap with the predecessor (e.g. in the phenomena it pertains to). However, this does not bear on the present line of argument.
14. See also Popper (2002 [1963], p. 293): 'even before a theory has ever undergone an empirical test we may be able to say whether, provided it passes certain specified tests, it would be an improvement on other theories with which we are acquainted.'
15. See also Salmon (1968) for an attempt to develop Reichenbach's view. I share Salmon's view that the proper solution to the problem of induction is pragmatic in nature, but disagree that the use of corroboration to guide decisions, discussed subsequently, amounts to an ampliative form of inference.
16. Admittedly, though, in that case the probability of its approximate truth might be unity on an appropriate construal of approximate truth.
17. Putnam (1969) advocates just such an inductive defence of induction.
18. See Ayer (1963), Gillies (2000, pp. 119–125), and Hájek (2007).
19. As an aside, it is interesting that Keynes (1921, p. 233) thought that: 'The object of increasing the number of instances arises out of the fact that we are nearly always aware of *some* difference between the instances, and that even where the known difference is insignificant we may suspect, especially when our knowledge of the instances is very incomplete, that there may be more . . . For this reason, and for this reason only, new instances are valuable.' Keynes nevertheless—and curiously, to my mind—believed that theories can be confirmed rather than merely corroborated.
20. In his words: 'Our point of view remains in all cases the same: *to show that there are rather profound psychological reasons which make the exact or approximate agreement that is observed between the opinions of different individuals very natural, but that there are no reasons, rational, positive, or metaphysical, that can give this fact any meaning beyond that of a simple agreement of subjective opinions*' (De Finetti 1937, p. 152).
21. Admittedly, Popper did not distinguish here between prior and posterior probabilities, as perhaps he should have. In short, to aim for theories with low prior probabilities is not to aim for theories with low posterior probabilities.

22. Note that probability zero should *not* be confused with impossibility, at least where infinities are concerned. See Williamson (2007).
23. For Keynes, the means by which the assignment is reached is also significant. (Otherwise, one could simply guess correctly and have a perfectly 'rational' degree of belief.) See Rowbottom (2008e).
24. Popper assumed that probabilities are standard real numbers; if we allow for infinitesimals, his argument would instead be that the probability of any universal hypothesis is infinitesimally small.
25. We will come back to this idea in Chapter 7.
26. Since it is so often missed, it is worth re-emphasising that Popper always denied the possibility of conclusive falsification. In his words (Popper 1959, p. 50): '[N]o conclusive disproof of a theory can ever be produced; for it is always possible to say that the experimental results are not reliable . . . '
27. The principle of indifference leads to several problems, such as Bertrand's paradox, which suggest that the logical interpretation is untenable. For the moment, however, we can put this to one side. Advocates of the logical interpretation *do* accept such a principle, or a similar one which leads to an identical probability assignment in this instance (e.g. the maximum entropy principle of Jaynes 1957). We will come back to this in the next chapter.
28. Put aside the thesis of Duhem (1954 [1906], p. 183) that 'an experiment . . . can never condemn an isolated hypothesis but only a whole theoretical group', for simplicity's sake, for the moment. The significance of this for Popper's account of corroboration is considered in Chapter 5.
29. Consider the subset of ordered quadruples of natural numbers (a, b, c, d), such that a and b are coprime, c and d are coprime, $a \geq 0$, $c \geq 0$, $b > 0$, and $d > 0$. Now let a/b represent r and c/d represent s. There is an injective function from the aforementioned subset to the set of natural numbers: (0, 1, 0, 1) maps onto 0; (1, 1, 0, 1) maps onto 1; (0, 1, 1, 1) maps onto 2; etc.
30. As Keynes (1921, p. 51) notes, however, the negative formulation thereof remains plausible: 'two propositions cannot be equally probable, so long as there is any ground for discriminating between them.' See Rowbottom (2008e).
31. See Sober (1975), for instance, who judges that it is not. I believe this goes for any of the ways in which simplicity can be understood, as detailed in Rowbottom (2009a).
32. For one of the best treatments of evidentialism available, see Conee and Feldman (2004).
33. As Keuth (2005, §5.52) illustrates, Popper suggested at some points that we should not believe in the truth, but only in the verisimilitude, of theories. I agree with Keuth (2005, §5.6) that this is a mistake, as illustrated by my subsequent discussion of van Fraassen's view that '*rationality is bridled irrationality*'.
34. Note that this means that the central term on the denominator will be defunct if $P(h,b)$ is zero, and the significance of this given Popper's argument that the logical probability of all universal theories is zero. For more on this, see Rowbottom (In Progress).
35. Amusingly, Arago only really *rediscovered* a phenomenon earlier noted by Maraldi. See Hecht (1998, p. 486).
36. See Miller (1994) and Gattei (2009), for instance.
37. As Watkins (1968, p. 65) put it: 'our methods of hypothesis-selection in practical life should be well-suited to our practical aims, just as our methods of hypothesis-selection in theoretical science should be well suited to our theoretical aims; and the two kinds of method may very well yield different answers in a particular case.'

And Salmon (1981, p. 119) was at pains: 'to emphasise that, even if we are entirely justified in letting such considerations [as corroboration value] determine our theoretical preferences, it is by no means obvious that we are justified in using them as the basis for our preferences among generalisations which are to be used for prediction in the practical decision-making context.'

38. Miller is perhaps a little generous to Popper here. On the same page, Popper wrote: 'we should *prefer* the best tested theory as a basis for action. In other words, there is no "absolute reliance"; but since we *have* to choose, it will be "rational" to choose the best tested theory. This will be "rational" in the most obvious sense of the word known to me: the best tested theory is the one which, in the light of our *critical discussion*, appears to be the best so far; and I do not know of anything more "rational" than a well-conducted critical discussion.'
39. Similarly, Popper (2002 [1963], p. 69) elsewhere used an evolutionary analogy to suggest that if we proceed by eliminating false theories, we will arrive at the most truthlike theory we have available: 'The critical attitude might be described as the result of a conscious attempt to make our theories, our conjectures, suffer in our stead in the struggle for the survival of the fittest. It gives us a chance to survive the elimination of an inadequate hypothesis—when a more dogmatic attitude would eliminate it by eliminating us . . . We thus obtain the fittest theory within our reach by elimination of those which are less fit. (By "fitness" I do not mean merely "usefulness" but truth . . .)'. As we will see in Chapter 7, one should therefore accept that the aim of science, for the anti-inductivist, should be something other than truth or verisimilitude.
40. For the record, 'we don't make mistakes in performing tests' may be satisfactorily replaced with 'tests are reliable'.
41. In any event, *p* does not follow from 'I cannot presently doubt that *p*', or even '*p* is undoubtable'.
42. See also Andersson (1994, p. 72).
43. Note well the difference between this hypothesis, which concerns the propensity in any given case, and one that concerns only the frequency of successes in the limit. This difference corresponds to the one between the approach of Neyman and the approach of Mayo and Spanos. If we only knew the frequency in the limit, for example, then we might not be justified in believing the ball was black in any particular case. However, we might be justified in sticking to the strategy of believing the ball to be black *because this would be successful in the long run.*
44. We would have to think that we can latch on to such processes in order for an evolutionary argument for the aim of science as truth to have *any* chance of going through. See the discussion in Chapter 7 and Rowbottom (2010).
45. Howson (2000, p. 3) also suggests that results from machine learning are philosophically significant.
46. This should be put in context, however, of Carnap's subsequent comments on the limitations of *deductive* machines. He suggests, for instance, that creativity is essential to solve problems.
47. Datteri et al. (2005) suggest that: '[A] familiar regress in epistemological discussions of induction arises as soon as one appeals to past performances of these systems in order to conclude that good showings are to be expected in their future outings as well.' However, it seems to me that this problem goes equally for assessing the probability of success of any process (and we normally consider this to be possible). We should not have one rule for (alleged) mechanical induction, and another for (alleged) gambling systems.

48. For more details, see Gillies (1996, 2003) and Muggleton et al. (1992).
49. This is true even if Datteri et al. (2005) are correct that:

> [T]he general notion of an autonomous learning system which, in addition to advancing learning hypotheses, is capable of retracting empirically inadequate hypotheses and revising its background knowledge base accordingly, does not require one to introduce notions of consequence that exceed or otherwise cannot be captured within the framework of deductive reasoning.

NOTES TO CHAPTER 3

1. Popper was a pluralist concerning probability, and held that the propensity interpretation is also applicable in some contexts. As will be discussed in the following, it is sometimes possible to interpret the probabilities in the corroboration function in this propensity sense, e.g. when considering statistical laws, although Popper did not suggest that we ought to do so.
2. It is true that the consequence is non-trivial in so far as such an observation does rule out some possible theories of light (given relevant auxiliary hypotheses). However, the experiment was seen to be a success for Fresnel's theory, rather than a failure for others.
3. I will not presently discuss whether the requirement for 'new evidence' is important, whether we should prefer theories that provide the best explanations (in anything more than a pragmatic sense), or whether prediction carries more weight than explanation. See the next chapter.
4. Popper (1959, app. *ix) did say that: '*there cannot be a metric of logical probability which is based upon purely logical considerations* . . . its metric depends upon our empirical knowledge'. However, as Rowbottom (2008e) shows, Keynes (1921) also thought that empirical constraints were significant; the point is simply that the probabilities are immutable once these constraints, which may be expressed as propositions, are taken into account. Let two different empirical constraints be E_1 and E_2. It is not the case that $P(e,h)$, say, will ever change depending on these constraints. But it is the case that $P(e,hE_1)$ and $P(e,hE_2)$ may have different *immutable* values. So, in short, a change in constraints just changes the conditional probabilities we're interested in. (After the aforementioned quotation, indeed, Popper added, 'These difficulties can be largely, but not entirely, overcome by making use of our "background knowledge"'.)
5. De Finetti (1972, p. 27) adds that: 'A measuring device is never exact and sure, unless it is imagined in an idealized version to be used in idealized conditions. In practice there are always, also in physics, imperfections and factors of disturbance.' In what follows, I endeavour to be mindful of this restriction.
6. Keynes also allows for non-numerical probabilities, but measurement (e.g. ordering) remains a serious problem in this case. Popper held that all probabilities must be numerical.
7. Ramsey (1926) refutes Keynes's appeal to straightforward intuition on some occasions; see also Gillies (2000, p. 52) and Rowbottom (2008e) for a more complete examination of Keynes's position especially in relation to the modern alternative of Objective Bayesianism.
8. Carnap (1962, p. 564) also has a similar problem in motivating the use of c^* rather than $c^\dagger$—that is, requiring equipossibility over structure descriptions rather than state descriptions—which he doesn't manage to solve except by stipulation.

9. See also van Fraassen (1989, p. 315). An objection to this line of argument has recently been raised by Bangu (2010), but is refuted by Rowbottom and Shackel (In Press).
10. To be more specific, *b* might be expressed as 'All men have a 0.92 propensity of being mortal, and Socrates is a man', such that it entails 'Socrates has a 0.92 propensity of being mortal'.
11. In fact, understandably, Gillies (1998, §2) suggests, '[The] approach could be called the "topping-up" version of the logical interpretation of probability.' He goes on to explain it in a similar way: 'The idea is to start with purely subjective degrees of belief. We then add one rationality constraint (coherence) to obtain the axioms of probability. However, this might be "topped-up" by further rationality constraints derived from logical or inductive intuition. Thus the choice of different probabilities allowed by the subjective theory would be narrowed down, and eventually it might be possible to get back to a single rational degree of belief as in the original logical theory.'
12. For an overview of different possible versions of the propensity theory, including the version I here allude to, see Gillies (2000, ch. 6 and 7).
13. It might be possible for some element in *b* to affect this reasoning, but let's imagine there is no such element.
14. Logical omniscience, on the part of the agent, is usually assumed in such discussions. So $D(\sim p,p) = 0$, $D(p,\sim p) = 0$, and so on.
15. So we may consider $P_{me}(p,qb)$ and $P_{rival}(p,qb)$. See De Finetti (1937, pp. 146–147).
16. See also Jarvie (2001, pp. 5–6), who shows that on Popper's view: 'Acquisition of knowledge is not possible without the interaction of social life'.
17. Bayesians usually do accept that a more informed decision is a better decision, as suggested by the so-called principle of total evidence, that 'the total evidence available must be taken as a basis' when a probability is calculated (Carnap 1962, p. 211). Recall also the result of Good (1967) and Ramsey (1990), mentioned in Chapter 1, that collecting 'free' evidence to make a decision, before making it, can never result in less expected utility that just making the decision without the evidence.
18. According to the subjective and logical views, we always have a well-defined set of background information. But if we assemble a number of people to discuss their views on some evidence given a theory, they will each offer different conditional probabilities, since they will each be using different background information. In short, we'll have $P(e,hb_1)$, $P(e,hb_2)$, . . . , $P(e,hb_n)$, where *n* is the number of participants. So what can we say is actually being evaluated? Here I assume it is the conditional probability on the union of the (appropriate and relevant) background information, although this might be problematic on closer analysis. Another option would be to suggest that we can arrive at some *B* which is fuzzy, yet nevertheless superior to any of b_1, . . . , b_n in so far as it is more pertinent.
19. Pritchard (2004, p. 339) also discusses a notion of epistemic entitlement precisely because he believes there are problems with showing how the acceptance of testimony could be justified.
20. Consider, for instance, the 'crucial encounter' (Cushing 1994, p. 118) between de Broglie and Pauli at the 1927 Solvay Conference. Both Bohm (1987, p. 39) and Cushing (1994, ch. 10) suggest that the Copenhagen theory might not have become the dominant one if de Broglie had fared better—for example, if he had been supported by Einstein.
21. It might seem curious to take this as a case for corroboration, given that one of the terms on the denominator of Popper's function is $P(eh,b) = P(e,hb)$ $P(h,b)$. In fact, this means that Popper's corroboration function must be

rejected. Yet his core desiderata can be retained, in a suitably modified corroboration function, as Gillies (1998) shows. For present purposes, it suffices to remember that the denominator is designed primarily to play a normalising role: the important measure is P(*e*,*hb*)-P(*e*,*b*).

22. Admittedly, it is possible to suggest that the aim of science is just to achieve theories with a particular virtue. But then this will still leave the problem of ranking the other virtues.

NOTES TO CHAPTER 4

1. Andersson (1994, ch. 6) discusses other constraints that apply, such as intersubjective testability.
2. Accidental observation of a black swan could therefore refute 'All swans are white', but not corroborate 'All swans are black'. Hence, corroboration is not just a surrogate for inductive confirmation. Note, however, that Popper (1959, §22) specifies that: 'a few stray basic statements contradicting a theory will hardly induce us to reject it as falsified. We shall take it as falsified only if we discover a *reproducible effect* which refutes the theory.' Thus the accidental observation would need to lead to discovery of a reproducible effect; in simple cases such as observing a black swan, this reproducibility will be evident. See also Popper (1983, p. 235).
3. Joseph Agassi has suggested to me, in personal correspondence, that 'Popper . . . rightly said that the initial degree of corroboration of a theory as it appears in public *prior to any test* has to be fairly high'. From a formal point of view, however, it appears that it could at best be zero. The crucial measure of corroboration, putting aside considerations relating to normalisation, is P(*e*,*hb*)-P(*e*,*b*). But this measure is only relevant if *e* is true, and we should not classify *e* as true unless we have made the relevant observations. Now consider a new hypothesis which hasn't been tested. Our prior evidence—and therefore any *e* which we can reasonably classify as true—is part of *b*. Therefore P(*e*,*hb*)-P(*e*,*b*)=0, provided *h* is consistent with *b*, unless we consider a hypothetical situation in which our background information does not contain all the evidence that we now have. Yet even if we do consider such a hypothetical scenario, it would only show whether the hypothesis nicely *accommodates* something we already know. Popper, on the other hand, advocated actual—rather than merely hypothetical—bold conjectures. See also the discussion in section 4.
4. Consider also: 'The appraisal of the corroboration . . . can be derived if we are given the theory as well as the accepted basic statements. It asserts the fact that these basic statements do not ontradict [*sic*] the theory, and it does this with due regard to the degree of testability of the theory, and to the severity of the tests to which the theory has been subjected, up to a stated period of time' (Popper 1959, p. 266).
5. If we just discuss increments of basic statements, rather than distinguish between test reports and background information, there is an even simpler objection. Let the statements in question concern the results of repeated spins of a roulette wheel. Today these may cast serious doubt on the hypothesis that the wheel is fair. Next week, however, they might cast no doubt whatsoever.
6. More carefully, this hypothesis might be framed in terms of propensities. We might suggest that the relevant experimental set-up (including me, crucially, in the example in the text) results in a 'heads' result with propensity 1/2 and a 'tails' result with propensity 1/2.

7. See also the discussion of Musgrave (1975) and Miller (1994, §2.2e).
8. Popper (1974b, p. 1080) appears to agree, although cautions that the situation is never so clean: 'in science we never mention explicitly more than the *relevant* evidence; the rest may appear as "background knowledge" . . . but only in a rough and implicit way . . . '
9. We need not really imagine ourselves in such scenarios, although this may be an aid to calculation. It should also be noted that in both cases the background knowledge is that of science as a whole (rather than any particular individual).
10. This sort of approach is discussed by Musgrave (1974a, §3), who also finds it wanting (although for different reasons). Musgrave (1974a, p. 19) instead prefers a 'variant [which] takes "background knowledge" to include only the best existing competing theory'. However, I would emphasise that corroboration relative to the best alternative to *h* at the time of its construction may not be the same as the corroboration relative to the best alternative to *h* at present. And it is difficult to see why we should be interested in the former, except to explain why we *once* preferred (or would have been right to prefer) *h*. The latter option remains a viable alternative to using $b^\dagger$, although not one that needs to be considered in depth in order to see the significance of the Big Test.
11. Achinstein (1994) makes a similar point concerning selection procedures, in response to Snyder (1994). In short, how *e* is arrived at is sometimes relevant to how (or even if) it bears on *h*.

NOTES TO CHAPTER 5

1. Earlier treatments include Dorling (1979) and Howson and Urbach (1989).
2. I follow Gillies (1993) in taking the Duhem and Quine theses to be distinct. The former is widely accepted as true, whereas the latter is not, which is why I focus on it here. For an interesting discussion of the background to the thesis, with reference to Poincaré as well as Duhem, see Brenner (1990).
3. See, however, Popper (1959, p. 42 and §20).
4. Popper (1959, p. 83) subsequently concedes that this cannot be correct for some auxiliary hypotheses, in particular '*singular* statements . . . [e.g.] the assumption that a certain observation or measurement which cannot be repeated may have been due to error.'
5. Leplin (1975, p. 345) also admits that 'it is difficult to fix a point at which the neutrino's theoretical utility became sufficient in strength and diversity to overcome its initial ad hoc character . . . '; so, in short, he considers it unclear when exactly it was no longer *ad hoc*. Unbeknownst to Leplin, Popper (1974e, p. 986) pre-empted him:

   > It is clear that, like everything in methodology, the distinction between an *ad hoc* hypothesis and a conservative auxiliary hypothesis is a little vague. Pauli introduced the hypothesis of the neutrino quite consciously as an *ad hoc* hypothesis. He had originally no hope that one day independent evidence would be found . . . So we have an example of an *ad hoc* hypothesis which, with the growth of knowledge, did shed its *ad hoc* character. And we have a warning here not to pronounce too severe an edict against *ad hoc* hypotheses: they may become testable after all, as may also happen to a metaphysical hypothesis. But in general, our criterion of testability warns us against *ad hoc* hypotheses . . .

6. These are: $r(h, e, b) = \log\frac{P(h,eb)}{P(h,b)}$ and $l(h, e, b) = \log\frac{P(e,hb)}{P(e,\sim hb)}$.

7. Indeed, Popper (1983, p. 236) notes, 'It may not always be easy to decide what we have to exclude from *b*', but unfortunately provides no further discussion of the issue.
8. Or to put it more carefully, following Gillies (1971, p. 235), 'We must require that the observed result . . . should have a small relative likelihood [according to the relevant hypothesis] but also that it should be *untypical* in this'.
9. Note, however, that 'effort' cannot be measured simply in terms of time, money, and resources. As such, the subsequent example is a peculiar case. A lot of time and money may have been spent testing a hypothesis but not severely, in which case the corroboration value may not be high.
10. Popper also explains 'ad hocness' in terms of independent testability at one point in his later work. He writes, 'I call a conjecture "*ad hoc*" if it is introduced . . . to explain a particular difficulty, but if . . . *it cannot be tested independently* . . . ' (1974e, p. 986).

NOTES TO CHAPTER 6

1. My own view is that this is somewhat too generous to Popper, although I agree wholeheartedly that the seeds of this idea are present in his work—in particular, if we think of parallels between his views in political philosophy and the philosophy of science, e.g. about how to organise institutions effectively to achieve peculiar ends—and agree unreservedly, as illustrated in Chapter 3, that he strongly emphasised 'the connection . . . between social interaction and the acquisition of knowledge' (Jarvie 2001, p. 5).
2. This text is based on a lecture originally delivered in 1953. Interestingly, the other place is in a footnote in Popper's revised version of 'What is Dialectic?', again in *Conjectures and Refutations*. In the original paper in *Mind*—Popper (1940)—the relevant footnote did not appear.
3. In fact, Kuhn (1970b, p. 233) suggested that 'the descriptive and the normative are inextricably mixed'.
4. Indeed, this fits nicely with the view of Bailey (2006) that Kuhn's model of normal science education is indoctrinatory.
5. For example, Gattei (2008, p. 40) notes that the provisional programme for the colloquium on which *Criticism and the Growth of Knowledge* was based: 'describes the session as follows:

July 13, Tuesday	*Criticism and the Growth of Knowledge I*
Chairman:	Sir Karl R. Popper
9:15–10	T.S. Kuhn: *Dogma versus Criticism*
10:15–11	P. Feyerabend: *Criticism versus Dogma*
11:15–12:45	Discussion'

   He adds: 'Fundamental here are the contrasting words "criticism" and "dogma", chosen in order to emphasize the differences and characterize the two opposing positions—two diametrically opposed positions.' Gattei (2008, p. 54) also defends the further claim that 'the critical reference of Kuhn's philosophy has always been Popper's falsificationism'.
6. Kuhn and Popper disagreed on the aim of science. However, this issue can be put to one side for present purposes.
7. For more on the sorts of activity supposed to occur in 'normal science', see Kuhn (1996 [1962], ch. 3) and Rowbottom (In Press D).
8. Kuhn does not mention Duhem, or Duhem's problem, explicitly; rather, I propose that this is a helpful way to understand his view that 'anomalous experiences may not be identified with falsifying ones' (Kuhn 1996 [1962], p. 146).

9. Writing of a similar earlier passage and anticipating some of this chapter's later findings, Musgrave (1974b, pp. 580–581) also notes that Popper's comments on dogmatism: '. . . might seem to conflict with his more frequent emphasis on the desirability of a critical attitude. The apparent conflict is heightened by the psychologicist terminology—to resolve it, we must read "attitude" in a non-psychological way in both places. But then what is of "considerable significance" is not a dogmatic attitude *as such*: a dogmatic attitude towards *T* will only be fruitful if it leads a scientist to improve *T*, to articulate and elaborate it so that it can deal with counterarguments; it will be unfruitful if it means merely that a scientist sticks to *T* without improving it.' I will later argue, however, that sometimes a dogmatic attitude may allow a scientist to do things that he otherwise would not.
10. I take it to be uncontroversial that this is a fair interpretation of Kuhn's view in the first edition of *The Structure of Scientific Revolutions*. Several commentators on this chapter have suggested it is a caricature of his later position. I disagree to the extent that I think he maintained that non-puzzle-solving functions are required only in extraordinary science, i.e. occasionally.
11. Of course, the *proper* mechanism by which scientists should come to see particular failures as indicating serious anomalies is never satisfactorily explained in Kuhn's work. And, furthermore, it appears that one scientist or another will have to start a chain reaction by questioning the boundaries of the disciplinary matrix. But for present purposes, let us put this to one side.
12. This may be slightly unfair to Kuhn, because puzzle solving involves a variety of activities. But Kuhn nowhere explained why some scientists might engage in one type of puzzle solving, and others engage in another. Furthermore, he did not endeavour to say how (or even if) a balance ought to be struck. For more on this issue, see Rowbottom (In Press D).
13. Bartley (1984, pp. 182–183) emphasised the importance of creativity, i.e. the imaginative function, as follows: '[A]n essential requirement is the *fertility* of the econiche: the econiche must be one in which the creation of positions and contexts, and the development of rationality, are truly inspired. Clumsily applied eradication of error may also eradicate fertility.' See also the discussion of evolutionary epistemology in the next chapter.
14. See also Rowbottom (In Press D), Bird (2000, pp. 68–69), and Hoyningen-Huene (1993).
15. An example may be "What are the best opening moves in chess?" It may be the case that some moves allow White to force a win, whereas others only allow drawing possibilities with best play, for example. We do not yet have computers sufficiently powerful to answer this question, although we know that in draughts 'perfect play by both sides leads to a draw' (Shaeffer et al. 2007, p. 1518).
16. The charge could equally be made that the critical function, in Popper's model, is composed of other functions. I accept this, however, as will be made apparent in the following.
17. As Settle et al. (1974, p. 89) note, pre-empting some of the ideas discussed here: 'objectivity is enhanced by the scientist who criticizes his own beliefs, but not undermined by one who criticizes only those of others, even if he does so out of sheer dogmatism.'
18. An interesting objection to the picture shown in Figure 6.6, indeed, is "Why should each and every scientist not perform all of *those* functions?" This was made to me when I presented this idea to an audience at the Future of Humanity Institute, University of Oxford.
19. In fact, as I argue in Rowbottom (In Press D), appeal to something like stances is necessary for Kuhn irrespective of whether any concessions are

made on the issue of criticism. In particular, appeal to stances can explain how different activities occur within a disciplinary matrix, i.e. account for the differences in puzzle solving activity. For example, they can explain why one scientist endeavours to articulate a theory, whereas another seeks only to apply it in unproblematic circumstances, or why two different scientists look to different exemplars (and therefore puzzle solving strategies) to tackle one and the same puzzle.

20. Incidentally, there are variants of chess, such as Crazyhouse chess, where this sort of move is possible.
21. I say this in part because one commentator on the ideas in this chapter, who works in the 'logical' tradition, reacted by declaring that "Kuhn was [just] a sociologist". Not only is this wrong—as Jones (1986) shows—but also remarkably myopic.

NOTES TO CHAPTER 7

1. This should not be confused with an alternative programme that may be described similarly, the so-called EEM (Evolution of Epistemological Mechanisms) programme, which involves 'extension of the biological theory of evolution to those aspects or traits of animals which are the biological substrates of cognitive activity' (Bradie 1990, pp. 245–246).
2. It should be noted that this definition requires interpretation in order for the analogy with competitive games, like chess and football, to hold. In these cases, the probability is high that *one opponent or the other* will succeed in achieving the aim of the game (although the situation is complicated somewhat by the possibility of draws).
3. More carefully, we need only allow that there is at least one way to do science—one method—such that it reliably makes continual progress towards *x*.
4. See Okasha (2006, p. 22) for a full derivation.
5. Bradie (1986, 1990), for instance, writes of the EET (Evolution of Epistemic Theories) programme.
6. We shall not consider Bird's (2007) recent proposal that scientific progress is best understood in terms of knowledge. This is not only due to the reservations of Rowbottom (2008d), but also because knowledge requires truth.
7. The absence of cumulativity in this case may lead some to think that the analogy with evolution is inappropriate. I am not so sure. Imagine a small population of beings that reproduce, and have offspring that never manage to reproduce because they are all wiped out by a terrible virus before reaching sexual maturity. Clearly there may have been selection from the point of view of Price's equation. Some of the small initial population may have been killed by predators because of unfavourable traits. In that event, fewer of the offspring may possess those traits than they would have in the absence of the predators.
8. It is unclear, in any event, why the entities in P must not be of different types *provided* that they each share the relevant attribute (and can appropriately 'reproduce').
9. According to the result of Miller (1974), for instance, the number of true members would be equal to the number of false members.
10. Compare this earlier passage, which may appear to be in conflict: 'Experiences can motivate a decision . . . but a basic statement cannot be justified by them—no more than by thumping the table' (Popper 1959, p. 105). For further discussion, see Musgrave (2009).
11. Just a little later, interestingly, Popper (1981, p. 83) advocated yet another, distinct, understanding of progress: 'the tentative adoption of a new conjecture or

theory may solve one or two problems, but it invariably opens up many *new* problems; for a new revolutionary theory functions exactly like a new and powerful sense organ. If the progress is significant then the new problems will differ from the old problems: the new problems will be on a radically different level of depth. This happened, for example, in relativity . . . in quantum mechanics . . . in molecular biology . . . This, I suggest, is the way in which science progresses.' First, however, the notion of 'depth' here is rather vague. (From what viewpoint should 'depth' be evaluated?) Second, it is unclear why we should expect falsification to result in new problems with radically different levels of depth more often than not; and Popper offers no argument that we should.

12. There are some rather obvious problems with van Fraassen's definition, e.g. 'All tables are composed of atoms' is a hypothesis concerning observables which might be empirically adequate, in the sense he intends, even if it is false. For present purposes, we shall assume that such difficulties can be dealt with.
13. See also the objection of Dicken (2007), which does not affect the argument here. Modal realism is not presupposed.
14. Admittedly this may also be the case for theories that are apparently empirically inadequate, due to Duhem's thesis. But claims concerning empirical adequacy encounter the special difficulty discussed later whereas claims concerning empirical inadequacy do not.
15. This might even count as a minor victory for van Fraassen, given his epistemological voluntarism. There is, at least, no reason to favour the view that the aim of science is truth *over* the view that the aim of science is empirical adequacy.

NOTES TO CHAPTER 8

1. I am, however, beginning to develop a novel 'cognitive instrumentalist' stance which avoids the need to appeal to the rather mysterious faculty of creativity that Popper posits. See Rowbottom (In Press B).
2. Agassi (2008, pp. 228–229) relates the following story:

   > Bartley had the temerity to accuse the philosopher [Popper] then of having surreptitiously changed his ground. Naturally, the philosopher heard about it. Naturally, he was furious. He demanded proof . . . Bartley provided some. Exhibit A was the classic, lovely, most impressive "What is Dialectic?" (1940) . . . The concluding assertion in the original essay is positivistic . . . and it is appropriately altered as reissued in the *Conjectures and Refutations* collection . . . The philosopher was told about Bartley's response; he was not lost for words, oh, no. He drew out his evidence and showed the emissaries his own copy of the reprint of the original paper . . . and it was found corrected by a fountain pen, clearly not by a ballpoint pen—thus proving it had been done very early, before Bartley and his likes appeared on the scene . . . But this is not the end of it. The story crossed the Atlantic and reached Boston, where I was teaching then . . . It rang a bell: I had seen pages of the collection once on the philosopher's table, resting there and awaiting last inspection on their way to the printer . . . I had chanced then upon the original ending of "What is Dialectic?" and drew the philosopher's attention to it. To my surprise, he saw nothing wrong with it. I directed his attention then to its positivistic character. He looked around for a pen. This was a contingency I was very well used to since my long days as his assistant; I automatically offered him mine. He

refused it. With some effort he finally found and used a familiar old fountain pen, the one with the thick nub that writes so very beautifully. He staunchly refused to tell me why he chose it in preference to a simple ballpoint pen.

On a similar note, Watkins once said to Popper in his seminar (according to Lakatos): 'Karl, you are dishonest. You hate criticism' (Lakatos and Feyerabend 1999, p. 189).

3. Popper was certainly not the first, of course, to suggest that being critical is important in some epistemological sense. As Cohen and Nagel (1934, ch. X) suggested, for example: 'What is called *scientific method* differs radically from these [other methods] by encouraging and developing the utmost possible doubt, so that what is left after such doubt is always supported by the best available evidence.' Nevertheless, Popper was the first to put criticism centre stage in a methodological respect, and to emphasise—with the possible exception of Bartley—that criticism may be construed as radically distinct from justification.
4. To reject scientific realism is not to embrace instrumentalism, i.e. to reject the so-called semantic thesis of scientific realism (Psillos 1999) that scientific theories should be construed literally.
5. David Miller assures me that Popper and Bartley were perfectly aware of externalism, but that they ignored it.
6. There may be evidential reasons on the view of Williamson (2002), where internal justification is not required for knowledge and all knowledge is evidence.
7. Agassi (1973, p. 397) suggests that pragmatism is on the cards for critical rationalists anyway: '[T]he debate about rationality is indeed about the act of commitment, whether to rationality or to some other creed. This is so much so that one may wonder if it is not the case that, since the debate is practical—conducive to an act—and therefore fundamental, these two facts do not make all participants in it pragmatists of one sort or another. To support this one might notice that already in his *Logik der Forschung* of 1935 Popper defended the scientific venture on the ground of its fruitfulness.'
8. In the words of Popper (1970, p. 57):

   > [A]t any moment we are prisoners caught in the framework of our theories; our expectations; our past experiences; our language. But we are prisoners in a Pickwickian sense: if we try, we can break out of our framework at any time. Admittedly, we shall find ourselves again in a framework, but it will be a better and roomier one; and we can at any moment break out again.
9. Popper (1983) does, however, give an account of learning where hypothesis generation by imagination is central.
10. Freeman and Skolimowski (1974) draw some interesting parallels between Peirce and Popper.

References

Achinstein, P. 1994. 'Explanation v. Prediction: Which Carries More Weight?', in D. Hull, M. Forbes, and R. M. Burian (eds.), *PSA 1994 Vol. 2*. East Lansing, MI: Philosophy of Science Association, pp. 156–164.

Agassi, J. 1959. 'Corroboration versus Induction', *British Journal for the Philosophy of Science* 9, 311–317.

Agassi, J. 1973. 'Rationality and the *Tu Quoque* Argument', *Inquiry* 16, 395–406.

Agassi, J. 2008. *A Philosopher's Apprentice: In Karl Popper's Workshop*. Amsterdam: Rodopi.

Agassi, J., Jarvie, I., and Settle, T. 1971. 'The Grounds of Reason', *Philosophy* 46, 43–50.

Albert, H. 1985. *Treatise on Critical Reason*. Translated by M. V. Rorty. Princeton, NJ: Princeton University Press.

Allan, D. W., Weiss, M. A., and Ashby, N. 1985. 'Around-the-World Relativistic Sagnac Experiment', *Science* 228, 69–70.

Alston, W. P. 1989. *Epistemic Justification: Essays in the Theory of Knowledge*. Ithaca, NY: Cornell University Press.

Alston, W. P. 1996. *A Realist Conception of Truth*. Ithaca, NY: Cornell University Press.

Andersson, G. 1994. *Criticism and the History of Science: Kuhn's, Lakatos's and Feyerabend's Criticisms of Critical Rationalism*. Leiden: Brill.

Armstrong, D. M. 1973. *Belief, Truth and Knowledge*. Cambridge: Cambridge University Press.

Armstrong, D. M. 1983. *What Is a Law of Nature?* Cambridge: Cambridge University Press.

Artigas, M. 1999. *The Ethical Nature of Karl Popper's Theory of Knowledge*. Bern: Peter Lang.

Audi, R. 1994. 'Dispositional Beliefs and Dispositions to Believe', *Noûs* 28, 419–434.

Ayer, A. J. 1963. 'Two Notes on Probability', in *The Concept of a Person and Other Essays*. London: Macmillan, pp. 188–208.

Ayer, A. J. 1974. 'Truth, Verification, and Verisimilitude', in P. A. Schilpp (ed.), *The Philosophy of Karl Popper (Library of Living Philosophers XIV)*. La Salle, IL: Open Court, pp. 684–692.

Bacon, F. 1994. *Novum Organum*. Edited by P. Urbach and J. Gibson. Chicago: Open Court. (1st edition published in 1620.)

Bailey, R. 2006. 'Science, Normal Science and Science Education—Thomas Kuhn and Education', *Learning for Democracy* 2, 7–20.

Bangu, S. 2010. 'On Bertrand's Paradox', *Analysis* 70, 30–35.

Barnes, E. C. 2005. 'Predictivism for Pluralists', *British Journal for the Philosophy of Science* 56, 421–450.

Barratt, P. E. H. 1971. *Bases of Psychological Methods*. Queensland: John Wiley and Sons.
Bartley, W. W. 1962. *The Retreat to Commitment*. New York: Alfred A. Knopf, 1st edition.
Bartley, W. W. 1968. 'Theories of Demarcation between Science and Metaphysics', in I. Lakatos and A. Musgrave (eds.), *Problems in the Philosophy of Science*. Amsterdam: North-Holland, pp. 40–64.
Bartley, W. W. 1983. 'The Alleged Refutation of Pancritical Rationalism', in *Proceedings of the 11th International Conference on the Unity of the Sciences*. New York: ICF Press, pp. 1139–1179.
Bartley, W. W. 1984. *The Retreat to Commitment*. La Salle, IL: Open Court, 2nd edition.
Baumann, P. In Press. 'Empiricism, Stances and the Problem of Voluntarism', *Synthese*. (DOI: 10.1007/s11229-009-9519-7)
Bertrand, J. 1960. *Calcul des Probabilités*. New York: Chelsea, 3rd edition. (1st edition published in 1889.)
Bird, A. 2000. *Thomas Kuhn*. Chesham: Acumen.
Bird, A. 2001. 'Necessarily, Salt Dissolves in Water', *Analysis* 61, 267–274.
Bird, A. 2004. 'Thomas Kuhn', in E. N. Zalta (ed.), *The Stanford Encyclopedia of Philosophy*. URL: http://plato.stanford.edu/archives/fall2004/entries/thomas-kuhn/.
Bird, A. 2007. 'What is Scientific Progress?', *Noûs* 41, 64–89.
Bird, A. Forthcoming. 'Inductive Knowledge', in D. Pritchard (ed.), *Routledge Companion to Epistemology*. London: Routledge.
Bohm, D. 1987. 'Hidden Variables and the Implicate Order', in B. J. Hiley and D. F. Peat (eds.), *Quantum Implications: Essays in Honour of David Bohm*. London: Routledge, pp. 33–45.
BonJour, L. 1985. *The Structure of Empirical Knowledge*. Cambridge, MA: Harvard University Press.
Booth, A. R. 2006. 'Can There Be Epistemic Reasons for Action?', *Grazer Philosophische Studien* 73, 162–173.
Booth, A. R. 2007. 'The Two Faces of Evidentialism', *Erkenntnis* 67, 401–417.
Booth, A. R. 2009. 'Motivating Epistemic Reasons for Action', *Grazer Philosophische Studien* 78, 265–271.
Bradie, M. 1986. 'Assessing Evolutionary Epistemology', *Biology and Philosophy* 1, 401–459.
Bradie, M. 1990. 'The Evolution of Scientific Lineages', *PSA 1990 Vol. 2*, 245–254.
Brenner, A. A. 1990. 'Holism a Century Ago: The Elaboration of Duhem's Thesis', *Synthese* 73, 325–335.
Brush, S. G. 1989. 'Prediction and Theory-Evaluation: The Case of Light Bending', *Science* 246, 1124–1129.
Bueno, O. 1999. 'What Is Structural Empiricism? Scientific Change in an Empiricist Setting', *Erkenntnis* 50, 59–85.
Bueno, O. 2000. 'Empiricism, Scientific Change and Mathematical Change', *Studies in History and Philosophy of Science* 31, 269–296.
Bueno, O. 2008. 'Structural Realism, Scientific Change, and Partial Structures', *Studia Logica* 89, 213–235.
Campbell, D. T. 1974. 'Evolutionary Epistemology', in P. A. Schilpp (ed.), *The Philosophy of Karl Popper*. La Salle, IL: Open Court, pp. 413–463.
Carnap, R. 1962. *Logical Foundations of Probability*. Chicago: University of Chicago Press, 2nd edition.
Carnap, R. 1968. 'Inductive Intuition and Inductive Logic', in I. Lakatos (ed.), *The Problem of Inductive Logic (Studies in Logic and the Foundations of Mathematics Vol. 51)*. Amsterdam: North-Holland, pp. 258–314.

Cartwright, N. 1983. *How the Laws of Physics Lie*. Oxford: Oxford University Press.

Chakravartty, A. 2004. 'Stance Relativism: Empiricism versus Metaphysics', *Studies in History and Philosophy of Science* 35, 173–184.

Chakravartty, A. In Press. 'A Puzzle about Voluntarism about Rational Epistemic Stances', *Synthese*.

Chalmers, D. Forthcoming. *Constructing the World*.

Churchland, P. M. 1985. 'The Ontological Status of Observables: In Praise of Superempirical Virtues', in P. M. Churchland and C. A. Hooker (eds.), *Images of Science: Essays on Realism and Empiricism*. Chicago: University of Chicago Press, pp. 35–47.

Cohen, L. J. 1966. 'What has Confirmation to Do with Probabilities?', *Mind* 75, 463–481.

Cohen, L., Manion, L., and Morrison, K. 2003. *Research Methods in Education*. London: Routledge.

Cohen, M. R. 1953. *Reason and Nature: The Meaning of Scientific Method*. London: The Free Press of Glencoe, Collier-Macmillan. (1st edition published in 1931.)

Cohen, M. R., and Nagel, E. 1934. *An Introduction to Logic and Scientific Method*. London: Routledge and Kegan Paul.

Cohen, S. 2002. 'Basic Knowledge and the Problem of Easy Knowledge', *Philosophy and Phenomenological Research* 65, 309–329.

Cohen, S. 2005. 'Why Basic Knowledge is Easy Knowledge', *Philosophy and Phenomenological Research* 70, 417–430.

Conee, E., and Feldman, R. 2004. *Evidentialism: Essays in Epistemology*. Oxford: Oxford University Press.

Curd, M., and Cover, J. A. (eds.) 1998. *Philosophy of Science: The Central Issues*. New York: W. W. Norton.

Cushing, J. T. 1994. *Quantum Mechanics: Historical Contingency and the Copenhagen Hegemony*. Chicago: University of Chicago Press.

Datteri, E., Hosni, H., and Tamburrini, G. 2005. 'Machine Learning from Examples: A Non-Inductivist Analysis', *Logic and Philosophy of Science* 3.

David, M. 2001. 'Truth as the Epistemic Goal', in M. Steup (ed.), *Knowledge, Truth and Duty: Essays on Epistemic Justification, Responsibility and Virtue*. Oxford: Oxford University Press, pp. 151–169.

De Finetti, B. 1937. 'Foresight: Its Logical Laws, Its Subjective Sources', in H. E. Kyburg and H. E. Smokler (eds.), *Studies in Subjective Probability*. New York: Wiley, pp. 93–158.

De Finetti, B. 1972. 'Subjective or Objective Probability: Is the Dispute Undecidable?', *Symposia Mathematica* 9, 21–36.

Dicken, P. 2007. 'Constructive Empiricism and the Metaphysics of Modality', *British Journal for the Philosophy of Science* 58, 605–612.

Diller, A. 2008. 'Testimony from a Popperian Perspective', *Philosophy of the Social Sciences* 38, 419–456.

Domondon, A. T. 2009. 'Kuhn, Popper, and the Superconducting Supercollider', *Studies in History and Philosophy of Science* 40, 301–314.

Dorling, J. 1979. 'Bayesian Personalism, the Methodology of Scientific Research Programmes, and Duhem's Problem', *Studies in History and Philosophy of Science* 10, 177–187.

Dretske, F. 1977. 'Laws of Nature', *Philosophy of Science* 44, 248–268.

Duhem, P. M. M. 1954. *The Aim and Structure of Physical Theory*. Translated by P. P. Wiener. Princeton, NJ: Princeton University Press. (1st English edition published in 1906.)

Ellis, B., and Lierse, C. 1994. 'Dispositional Essentialism', *Australasian Journal of Philosophy* 72, 27–45.

Fitelson, B., and Waterman, A. 2005. 'Bayesian Confirmation and Auxiliary Hypotheses Revisited: A Reply to Strevens', *British Journal for the Philosophy of Science* 56, 293–302.

Fitelson, B., and Waterman, A. 2007. 'Comparative Bayesian Confirmation and the Quine-Duhem Problem: A Rejoinder to Strevens', *British Journal for the Philosophy of Science* 58, 333–338.

Foley, R. 1987. *The Theory of Epistemic Rationality*. Cambridge, MA: Harvard University Press.

Foley, R. 1991. 'Evidence and Reasons for Belief', *Analysis* 51, 98–102.

Fox, M. F. 1994. 'Scientific Misconduct and Editorial and Peer Review Processes', *The Journal of Higher Education* 65, 298–309.

Freeman, E., and Skolimowski, H. 1974. 'The Search for Objectivity in Peirce and Popper', in P. A. Schilpp (ed.), *The Philosophy of Karl Popper*. La Salle, IL: Open Court, pp. 464–519.

French, S., and Kamminga, H. 1993. *Correspondence, Invariance and Heuristics: Essays in Honour of Heinz Post (Boston Studies in the Philosophy of Science)*. Dordrecht: Kluwer.

Fuller, S. 2003. *Kuhn vs. Popper: The Struggle for the Soul of Science*. London: Icon Books.

Gattei, S. 2008. *Thomas Kuhn's "Linguistic Turn" and the Legacy of Logical Empiricism*. Aldershot: Ashgate.

Gattei, S. 2009. *Karl Popper's Philosophy of Science: Rationality without Foundations*. London: Routledge.

Giles, J. 2005. 'Internet Encyclopaedias Go Head to Head', *Nature* 438, 900–901.

Gillies, D. A. 1971. 'A Falsifying Rule for Probability Statements', *British Journal for the Philosophy of Science* 22, 231–261.

Gillies, D. A. 1991. 'Intersubjective Probability and Confirmation Theory', *British Journal for the Philosophy of Science* 42, 513–533.

Gillies, D. 1993. *Philosophy of Science in the Twentieth Century: Four Central Themes*. Oxford: Blackwell.

Gillies, D. A. 1996. *Artificial Intelligence and Scientific Method*. Oxford: Oxford University Press.

Gillies, D. A. 1998. 'Confirmation Theory', in D. M. Gabbay and P. Smets (eds.), *Handbook of Defeasible Reasoning and Uncertainty Management Systems, Vol. 1*. Dordrecht: Kluwer, pp. 135–167.

Gillies, D. A. 2000. *Philosophical Theories of Probability*. London: Routledge.

Gillies, D. A. 2003. 'The Problem of Induction and Artificial Intelligence', presented at Karl R. Popper: A Revision of his Legacy conference, La Coruña, Spain.

Glymour, C. 1980. *Theory and Evidence*. Princeton, NJ: Princeton University Press.

Goldman, A. I. 1967. 'A Causal Theory of Knowing', *Journal of Philosophy* 64, 357–372.

Goldman, A. I. 1975. 'Innate Knowledge', in S. P. Stich (ed.), *Innate Ideas*. Berkeley: University of California Press, pp. 111–120.

Goldman, A. I. 1979. 'What Is Justified Belief?', in G. Pappas (ed.), *Justification and Knowledge*. Dordrecht: Reidel, pp. 1–23.

Good, I. J. 1967. 'On the Principle of Total Evidence', *British Journal for the Philosophy of Science* 17, 319–321.

Gower, B. 1997. *Scientific Method: An Historical and Philosophical Introduction*. London: Routledge.

Greco, J. 2008. 'Virtue Epistemology', in E. N. Zalta (ed.), *The Stanford Encyclopedia of Philosophy*. URL: http://plato.stanford.edu/archives/fall2008/entries/epistemology-virtue/.

Grünbaum, A. 1976. 'Ad Hoc Auxiliary Hypotheses and Falsificationism', *British Journal for the Philosophy of Science* 27, 329–362.

Haack, S. 1991. 'What is "The Problem of the Empirical Basis", and Does Johnny Wideawake Solve It?', *British Journal for the Philosophy of Science* 42, 369–389.

Hacking, I. 1975. *The Emergence of Probability.* Cambridge: Cambridge University Press.

Hafele, J. C., and Keating, R. E. 1972. 'Around-the-World Atomic Clocks: Observed Relativistic Time Gains', *Science* 177, 168–170.

Hájek, A. 2003. 'What Conditional Probability Could Not Be', *Synthese* 137, 273–323.

Hájek, A. 2005. 'Scotching Dutch Books?', *Philosophical Perspectives* 19, 139–151.

Hájek, A. 2007. 'The Reference Class Problem is Your Problem Too', *Synthese* 156, 563–585.

Harker, D. 2008. 'On the Predilections for Predictions', *British Journal for the Philosophy of Science* 59, 429–453.

Hauptli, B. W. 1991. 'A Dilemma for Bartley's Pancritical Rationalism', *Philosophy of the Social Sciences* 21, 86–89.

Hecht, E. 1998. *Optics.* Reading, MA: Addison-Wesley, 3rd edition.

Helm, P. 1987. 'On Pan-Critical Irrationalism', *Analysis* 47, 24–28.

Hempel, C. G. 1965. *Aspects of Scientific Explanation and other Essays in the Philosophy of Science.* New York: Free Press.

Hendry, R. F., and Rowbottom, D. P. 2009. 'Dispositional Essentialism and the Necessity of Laws', *Analysis* 69, 668–677.

Howson, C. 1973. 'Must the Logical Probability of Laws Be Zero?', *British Journal for the Philosophy of Science* 24, 153–182.

Howson, C. 2000. *Hume's Problem: Induction and the Justification of Belief.* Oxford: Oxford University Press.

Howson, C., and Urbach, P. 1989. *Scientific Reasoning: The Bayesian Approach.* Chicago: Open Court.

Hoyningen-Huene, P. 1993. *Reconstructing Scientific Revolutions: Thomas S. Kuhn's Philosophy of Science.* Chicago: University of Chicago Press.

Huber, F. 2008. 'Milne's Argument for the Log-Ratio Measure', *Philosophy of Science* 75, 413–420.

Hudson, R. G. 2007. 'What's Really at Issue with Novel Predictions?', *Synthese* 115, 1–20.

Hull, D. L. 1988. *Science as a Process: An Evolutionary Account of the Social and Conceptual Development of Science.* Chicago: University of Chicago Press.

Jäger, C. 2007. 'Is Coherentism Coherent?', *Analysis* 67, 341–344.

Jarvie, I. 2001. *The Republic of Science: The Emergence of Popper's Social View of Science 1935–1945.* Amsterdam: Rodopi.

Jaynes, E. T. 1957. 'Information Theory and Statistical Mechanics', *Physical Review* 106, 620–630.

Jaynes, E. T. 1973. 'The Well Posed Problem', *Foundations of Physics* 4, 477–492.

Jaynes, E. T. 2003. *Probability Theory: The Logic of Science.* Cambridge: Cambridge University Press.

Jones, K. 1986. 'Is Kuhn a Sociologist?', *British Journal for the Philosophy of Science* 37, 443–452.

Jones, W. E. (ed.) 2003. Special Issue on Controlling Belief, *The Monist* 85, 3.

Keene, G. B. 1961. 'Confirmation and Corroboration', *Mind* 70, 85–87.

Kekes, J. 1971. 'Watkins on Rationalism', *Philosophy* 46, 51–53.

Keuth, H. 2005. *The Philosophy of Karl Popper.* Cambridge: Cambridge University Press.
Keynes, J. M. 1921. *A Treatise on Probability.* London: Macmillan.
Kitcher, P. 1990. 'The Division of Cognitive Labor', *Journal of Philosophy* 87, 5–22.
Koop, T., and Pöschl, U. 2006. 'Systems: An Open, Two-Stage Peer Review Journal', *Nature.* URL: http://www.nature.com/nature/peerreview/debate/nature04988.html.
Kuhn, T. S. 1961. 'The Function of Dogma in Scientific Research', in A. C. Crombie (ed.), *Scientific Change.* New York: Basic Books, pp. 347–369.
Kuhn, T. S. 1970a. 'Logic of Discovery or Psychology of Research?', in I. Lakatos and A. Musgrave (eds.), *Criticism and the Growth of Knowledge.* Cambridge: Cambridge University Press, pp. 1–23.
Kuhn, T. S. 1970b. 'Reflections on My Critics', in I. Lakatos and A. Musgrave (eds.), *Criticism and the Growth of Knowledge.* Cambridge: Cambridge University Press, pp. 231–278.
Kuhn, T. S. 1977. *The Essential Tension: Selected Studies in Scientific Tradition and Change.* Chicago: University of Chicago Press.
Kuhn, T. S. 1996. *The Structure of Scientific Revolutions.* Chicago: University of Chicago Press, 3rd edition. (1st edition published in 1962.)
Lackey, J., and Sosa, E. (eds.) 2006. *The Epistemology of Testimony.* Oxford: Oxford University Press.
Ladyman, J. 2000. 'What's Really Wrong with Constructive Empiricism?', *British Journal for the Philosophy of Science* 51, 837–856.
Ladyman, J. 2009. 'Structural Realism', in E. N. Zalta (ed.), *The Stanford Encyclopedia of Philosophy.* URL: http://plato.stanford.edu/archives/sum2009/entries/structural-realism/.
Lakatos, I. 1974. 'Popper on Demarcation and Induction', in P. A. Schilpp (ed.), *The Philosophy of Karl Popper.* La Salle, IL: Open Court, pp. 241–273.
Lakatos, I., and Feyerabend, P. 1999. *For and Against Method: Including Lakatos's Lectures on Scientific Method and the Lakatos-Feyerabend Correspondence.* Edited by M. Motterlini. Chicago: University of Chicago Press.
Laudan, L. 1981. 'A Confutation of Convergent Realism', *Philosophy of Science* 48, 19–49.
Laudan, L. 1984. *Science and Values: The Aims of Science and Their Role in Scientific Debate.* Ewing: University of California Press.
Laudan, L. 1990. 'Aim-Less Epistemology?', *Studies in History and Philosophy of Science* 21, 315–322.
Leplin, J. 1975. 'The Concept of an Ad Hoc Hypothesis', *Studies in History and Philosophy of Science* 5, 309–345.
Lipton, P. 1998. 'The Epistemology of Testimony', *Studies in History and Philosophy of Science* 29, 1–31.
Lipton, P. 2004. *Inference to the Best Explanation.* London: Routledge.
Lorenz, K. 1977. *Behind the Mirror.* London: Methuen.
Lowe, E. J. 1998. *The Possibility of Metaphysics: Substance, Identity and Time.* Oxford: Oxford University Press.
Lowe, E. J. 2006. *The Four-Category Ontology: A Metaphysical Foundation for Natural Science.* Oxford: Oxford University Press.
Lowe, E. J. In Press. 'The Rationality of Metaphysics', *Synthese.* (DOI: 10.1007/s11229–009–9514-z)
Mayo, D. G., and Spanos, A. 2006: 'Severe Testing as a Basic Concept in a Neyman-Pearson Philosophy of Induction', *British Journal for the Philosophy of Science* 57, 323–357.

Miller, D. W. 1974. 'Popper's Qualitative Theory of Verisimilitude', *British Journal for the Philosophy of Science* 25, 166–177.

Miller, D. W. 1994. *Critical Rationalism: A Restatement and Defence.* La Salle, IL: Open Court.

Miller, D. W. Forthcoming. 'Deductivist Decision Making'.

Milne, P. 1996. 'Log[P(h/eb)/P(h/b)] Is the One True Measure of Confirmation', *Philosophy of Science* 63, 21–26.

Monton, B. (ed.) 2007. *Images of Empiricism.* Oxford: Clarendon Press.

Muggleton, S., King, R. D., and Sternberg, M. J. E. 1992. 'Protein Secondary Structure Prediction Using Logic-Based Machine Learning', *Protein Engineering* 5, 647–657.

Mumford, S. 1998. *Dispositions.* Oxford: Oxford University Press.

Mumford, S. 2004. *Laws in Nature.* London: Routledge.

Musgrave, A. 1974a. 'Logical versus Historical Theories of Confirmation', *British Journal for the Philosophy of Science* 25, 1–23.

Musgrave, A. 1974b. 'The Objectivism of Popper's Epistemology', in P. A. Schilpp (ed.), *The Philosophy of Karl Popper.* La Salle, IL: Open Court, pp. 560–596.

Musgrave, A. 1975. 'Popper and "Diminishing Returns from Repeated Tests"', *Australasian Journal of Philosophy* 53, 248–253.

Musgrave, A. 2009. 'Experience and Perceptual Belief', in Z. Parusniková and R. S. Cohen (eds.), *Rethinking Popper.* Dordrecht: Springer, pp. 5–19.

Newton-Smith, W. 1981. *The Rationality of Science.* London: Routledge.

Neyman, J. 1957. 'Inductive Behavior as a Basic Concept of Philosophy of Science', *Revue d'Institute International de Statistique* 25, 7–22.

Neyman, J., and Pearson, E. S. 1933. 'On the Problem of the Most Efficient Tests of Statistical Hypotheses', *Philosophical Transactions of the Royal Society A* 231, 289–337.

Niiniluoto, I. 1998. 'Verisimilitude: The Third Period', *British Journal for the Philosophy of Science* 49, 1–29.

Nilsson, J. 2006. 'On the Idea of Logical Presuppositions of Rational Criticism', in I. Jarvie, K. Milford, and D. Miller (eds.), *Karl Popper: A Centenary Assessment, Volume II: Metaphysics and Epistemology.* Aldershot: Ashgate, pp. 109–117.

Nola, R. 2005. 'Review of H. Keuth's "The Philosophy of Karl Popper"', *Notre Dame Philosophical Reviews.* URL: http://ndpr.nd.edu/review.cfm?id=4201.

Nozick, R. 1981. *Philosophical Explanations.* Cambridge, MA: Harvard University Press.

O'Hear, A. 1975. 'Rationality of Action and Theory-Testing in Popper', *Mind* 84, 273–276.

O'Hear, A. 1980. *Karl Popper.* London: Routledge and Kegan Paul.

Okasha, S. 2006. *Evolution and the Levels of Selection.* Oxford: Oxford University Press.

Owens, D. 2000. *Reason without Freedom: The Problem of Epistemic Normativity.* London: Routledge.

Pettit, P. N. 2006. 'When to Defer to Majority Testimony—and When Not', *Analysis* 66, 179–187.

Popper, K. R. 1940. 'What is Dialectic?', *Mind* 49, 402–436.

Popper, K. R. 1959. *The Logic of Scientific Discovery.* New York: Basic Books.

Popper, K. R. 1960. *The Poverty of Historicism.* London: Routledge and Kegan Paul, 2nd edition.

Popper, K. R. 1968. 'Remarks on the Problems of Demarcation and of Rationality', in I. Lakatos and A. Musgrave (eds.), *Problems in the Philosophy of Science.* Amsterdam: North-Holland, pp. 88–102.

Popper, K. R. 1970. 'Normal Science and Its Dangers', in I. Lakatos and A. Musgrave (eds.), *Criticism and the Growth of Knowledge*. Cambridge: Cambridge University Press, pp. 51–58.
Popper, K. R. 1972. *Objective Knowledge: An Evolutionary Approach*. Oxford: Oxford University Press.
Popper, K. R. 1974a. 'Ayer on Empiricism', in P. A. Schilpp (ed.), *The Philosophy of Karl Popper*. La Salle, IL: Open Court, pp. 1100–1114.
Popper, K. R. 1974b. 'Musgrave on Psychologism', in P. A. Schilpp (ed.), *The Philosophy of Karl Popper*. La Salle, IL: Open Court, pp. 1078–1080.
Popper, K. R. 1974c. 'The Psychological and Pragmatic Problems of Induction', in P. A. Schilpp (ed.), *The Philosophy of Karl Popper*. La Salle, IL: Open Court, pp. 1023–1027.
Popper, K. R. 1974d. 'Putnam on "Auxiliary Sentences", Called by Me "Initial Conditions"', in P. A. Schilpp (ed.), *The Philosophy of Karl Popper*. La Salle, IL: Open Court, pp. 993–999.
Popper, K. R. 1974e. '*Ad Hoc* Hypotheses and Auxiliary Hypotheses. The Falsifiability of Newton's Theory', in P. A. Schlipp (ed.), *The Philosophy of Karl Popper*. La Salle, IL: Open Court, pp. 986–987.
Popper, K. R. 1981. 'The Rationality of Scientific Revolutions', in I. Hacking (ed.), *Scientific Revolutions*. Oxford: Oxford University Press, pp. 80–106.
Popper, K. R. 1983. *Realism and the Aim of Science*. London: Routledge.
Popper, K. R. 1984. 'Evolutionary Epistemology', in J. W. Pollard (ed.), *Evolutionary Theory: Paths into the Future*. London: John Wiley and Sons, pp. 239–255.
Popper, K. R. 1999. *The Logic of Scientific Discovery*. London: Routledge, reprint edition.
Popper, K. R. 2002. *Conjectures and Refutations*. London: Routledge, classics edition. (1st edition published in 1963.)
Popper, K. R. 2003. *The Open Society and Its Enemies*. London: Routledge, classics edition. (1st edition published in 1945.)
Post, H. R. 1971. 'Correspondence, Invariance and Heuristics: In Praise of Conservative Induction', *Studies in History and Philosophy of Science* 2, 213–255.
Post, J. F. 1972. 'Paradox in Critical Rationalism and Related Theories', *Philosophical Forum* 3, 27–61.
Post, J. F. 1983. 'A Godëlian Theorem for Theories of Rationality', in *Proceedings of the 11th International Conference on the Unity of the Sciences*. New York: ICF Press, pp. 1071–1086.
Price, G. R. 1972. 'Extension of Covariance Selection Mathematics', *Annals of Human Genetics* 35, 485–490.
Pritchard, D. 2004. 'The Epistemology of Testimony', *Philosophical Issues* 14, 326–348.
Pritchard, D. 2009. 'Defusing Epistemic Relativism', *Synthese* 166, 397–412.
Pritchard, D. Forthcoming. 'Epistemic Relativism, Epistemic Incommensurability and Wittgensteinian Epistemology', in S. Hales (ed.), *The Blackwell Companion to Relativism*. Oxford: Blackwell.
Psillos, S. 1999. *Scientific Realism: How Science Tracks Truth*. London: Routledge.
Putnam, H. 1969. 'The "Corroboration" of Theories', in P. A. Schilpp (ed.) *The Philosophy of Karl Popper*. La Salle, IL: Open Court, pp. 221–240.
Quine, W. V. O. 1951. 'Two Dogmas of Empiricism', *Philosophical Review* 60, 20–43.
Radder, H. 1991. 'Heuristics and the Generalized Correspondence Principle', *British Journal for the Philosophy of Science* 42, 195–226.
Ramsey, F. P. 1926. 'Truth and Probability', in H. E. Kyburg and H. E. Smokler (eds.), *Studies in Subjective Probability*. New York: Wiley, pp. 61–92.
Ramsey, F. P. 1990. 'Weight or the Value of Knowledge', *British Journal for the Philosophy of Science* 41, 1–4.

Reichenbach, H. 1938. *Experience and Prediction*. Chicago: University of Chicago Press.
Rescher, N. 1987. *Scientific Realism: A Critical Reappraisal*. Dordrecht: D. Reidel.
Rice, S. 2004. *Evolutionary Theory*. Sunderland, MA: Sinauer.
Richmond, S. 1971. 'Can a Rationalist Be Rational about His Rationalism?', *Philosophy* 46, 54–55.
Rinard, P. M. 1976. 'Large-scale Diffraction Patterns from Circular Objects', *American Journal of Physics* 44, 70–76.
Rowbottom, D. P. 2004. 'Destructive Realism: Metaphysics as the Foundation of Natural Science', unpublished PhD thesis, Durham University.
Rowbottom, D. P. 2005. 'The Empirical Stance vs. the Critical Attitude', *South African Journal of Philosophy* 24, 200–223.
Rowbottom, D. P. 2006a. 'In Defence of Popper on the Logical Possibility of Universal Laws: A Reply to Contessa', *Philosophical Writings* 31, 53–60.
Rowbottom, D. P. 2006b. 'Kuhn versus Popper on Science Education: A Response to Richard Bailey', *Learning for Democracy* 2, 45–52.
Rowbottom, D. P. 2007a. '"In-Between Believing" and Degrees of Belief', *Teorema* 26, 131–137.
Rowbottom, D. P. 2007b. 'The Insufficiency of the Dutch Book Argument', *Studia Logica* 87, 65–71.
Rowbottom, D. P. 2007c. 'A Refutation of Foundationalism?', *Analysis* 67, 345–346.
Rowbottom, D. P. 2008a. 'An Alternative Account of Epistemic Reasons for Action: In Response to Booth', *Grazer Philosophische Studien* 76, 191–198.
Rowbottom, D. P. 2008b. 'The Big Test of Corroboration', *International Studies in Philosophy of Science* 22, 293–302.
Rowbottom, D. P. 2008c. 'Intersubjective Corroboration', *Studies in History and Philosophy of Science* 39, 124–132.
Rowbottom, D. P. 2008d. 'N-rays and the Semantic View of Scientific Progress', *Studies in History and Philosophy of Science* 39, 277–278.
Rowbottom, D. P. 2008e. 'On the Proximity of the Logical and "Objective Bayesian" Interpretations of Probability', *Erkenntnis* 69, 335–349.
Rowbottom, D. P. 2009a. 'Models in Biology and Physics: What's the Difference?', *Foundations of Science* 14, 281–294.
Rowbottom, D. P. 2009b. 'No Dilemma for Pancritical Rationalism: In Response to Hauptli', *Philosophy of the Social Sciences* 39, 490–494.
Rowbottom, D. P. 2010. 'Evolutionary Epistemology and the Aim of Science', *Australasian Journal of Philosophy* 88, 209–225.
Rowbottom, D. P. In Press A. 'Corroboration and Auxiliary Hypotheses: Duhem's Thesis Revisited', *Synthese*. (DOI: 10.1007/s11229–009–9643–4)
Rowbottom, D. P. In Press B. 'The Instrumentalist's New Clothes', *Philosophy of Science*.
Rowbottom, D. P. In Press C. 'Kuhn vs. Popper on Criticism and Dogmatism in Science: A Resolution at the Group Level', *Studies in History and Philosophy of Science*.
Rowbottom, D. P. In Press D. 'Stances and Paradigms: A Reflection', *Synthese*. (DOI: 10.1007/s11229–009–9524-x)
Rowbottom, D. P. In Progress. 'Popper's Measure of Corroboration and P(*h, b*)'.
Rowbottom, D. P., and Bueno, O. 2009. 'Why Advocate Pancritical Rationalism?', in Z. Parusniková and R. S. Cohen (eds.), *Rethinking Popper (Boston Studies in the Philosophy of Science, Vol. 272)*. Dordrecht: Springer, pp. 81–89.
Rowbottom, D. P., and Bueno, O. In Press A. 'How to Change It: Modes of Engagement, Rationality, and Stance Voluntarism', *Synthese*. (DOI: 10.1007/s11229–009–9521–0)

Rowbottom, D. P., and Bueno, O. In Press B. Special Issue of *Synthese* on Stance and Rationality.
Rowbottom, D. P., and Shackel, N. In Press. 'Bangu's Random Thoughts on Bertrand's Paradox', *Analysis*.
Russell, B. 1912. *The Problems of Philosophy*. London: Oxford University Press.
Salmon, W. C. 1965. 'The Concept of Inductive Evidence', *American Philosophical Quarterly* 2, 265–280.
Salmon, W. C. 1968. 'The Justification of Inductive Rules of Inference', in I. Lakatos (ed.), *The Problem of Inductive Logic* (*Studies in Logic and the Foundations of Mathematics Vol. 51*). Amsterdam: North-Holland, pp. 24–43.
Salmon, W. C. 1981. 'Rational Prediction', *British Journal for the Philosophy of Science* 32, 115–125.
Salmon, W. C. 1990. 'Rationality and Objectivity in Science (or Tom Kuhn Meets Tom Bayes)', in C. W. Savage (ed.), *Scientific Theories (Minnesota Studies in the Philosophy of Science XIV)*. Minneapolis: University of Minnesota Press, pp. 175–204.
Sankey, H. 2000. 'Methodological Pluralism, Normative Naturalism and the Realist Aim of Science', in R. Nola and H. Sankey (eds.), *After Popper, Kuhn and Feyerabend: Recent Issues in Theory of Scientific Method*. Dordrecht: Kluwer, pp. 211–229.
Scerri, E. R., and Worrall, J. 2001. 'Prediction and the Periodic Table', *Studies in History and Philosophy of Science* 32, 407–452.
Schwitzgebel, E. 2001. 'In-Between Believing', *Philosophical Quarterly* 51, 76–82.
Schwitzgebel, E. Forthcoming. 'Acting Contrary to Our Professed Beliefs, or the Gulf between Occurrent Judgment and Dispositional Belief'. URL: http://www.faculty.ucr.edu/~eschwitz/SchwitzAbs/ActBel.htm.
Settle, T. W. 1990. 'Swann versus Popper on Induction: An Arbitration', *British Journal for the Philosophy of Science* 41, 401–405.
Settle, T., Jarvie, I. C., and Agassi, J. 1974. 'Towards a Theory of Openness to Criticism', *Philosophy of the Social Sciences* 4, 83–90.
Shackel, N. 2007. 'Bertrand's Paradox and the Principle of Indifference', *Philosophy of Science* 74, 150–175.
Shackel, N. 2008. 'Coherentism and the Symmetry of Epistemic Support', *Analysis* 68, 226–234.
Shackel, N., and Rowbottom, D. P. In Progress. 'Objective Bayesianism and Bertrand's Paradox'.
Shaeffer, J., et al. 2007. 'Checkers is Solved', *Science* 317, 1518–1522.
Shearmur, J. 1996. *The Political Thought of Karl Popper*. London: Routledge.
Snyder, L. J. 1994. 'Is Evidence Historical?', in P. Achinstein and L. J. Snyder (eds.), *Scientific Methods: Conceptual and Historical Problems*. Malabar, FL: Krieger Publishing Company, pp. 95–117. Reprinted in Curd, M. and Cover, J. A. (eds.) 1998. *Philosophy of Science: The Central Issues*. New York: W. W. Norton, pp. 460–480.
Sober, E. 1975. *Simplicity*. Oxford: Oxford University Press.
Sosa, E., 1980. 'The Raft and the Pyramid: Coherence versus Foundations in the Theory of Knowledge', *Midwest Studies in Philosophy* 5, 3–25.
Steup, M. 2000. 'Doxastic Voluntarism and Epistemic Deontology', *Acta Analytica* 15, 25–56.
Steup, M. 2008. 'Doxastic Freedom', *Synthese* 161, 375–392.
Stoneham, T. 2007. 'A *Reductio* of Coherentism', *Analysis* 67, 254–257.
Strevens, M. 2001. 'The Bayesian Treatment of Auxiliary Hypotheses', *British Journal for the Philosophy of Science* 52, 515–537.
Strevens, M. 2003. 'The Role of the Priority Rule in Science', *Journal of Philosophy* 100, 55–79.

Strevens, M. 2005. 'The Bayesian Treatment of Auxiliary Hypotheses: Reply to Fitelson and Waterman', *British Journal for the Philosophy of Science* 56, 913–918.

Swann, J. 1999. 'What Happens When Learning Takes Place?', *Interchange* 30, 257–282.

Teller, P. 2004. 'What is a Stance?', *Philosophical Studies* 121, 159–170.

Tichý, P. 1974. 'On Popper's Definitions of Verisimilitude', *British Journal for the Philosophy of Science* 25, 155–160.

Toulmin, S. E. 1967. 'The Evolutionary Development of Natural Science', *American Scientist* 55, 456–471.

Van Fraassen, B. C. 1980. *The Scientific Image*. Oxford: Oxford University Press.

Van Fraassen, B. C. 1989. *Laws and Symmetry*. Oxford: Clarendon Press.

Van Fraassen, B. C. 2002. *The Empirical Stance*. New Haven, CT: Yale University Press.

Van Fraassen, B. C. 2004a. 'Précis of *The Empirical Stance*', *Philosophical Studies* 121, 127–132.

Van Fraassen, B. C. 2004b. 'Replies to Discussion on the Empirical Stance', *Philosophical Studies* 121, 171–192.

Van Fraassen, B. C. 2006. 'Structure: Its Shadow and Substance', *British Journal for the Philosophy of Science* 57, 275–307.

Van Fraassen, B. C. 2007. 'From a View of Science to a New Empiricism', in B. Monton (ed.), *Images of Empiricism*. Oxford: Clarendon Press, pp. 337–383.

Van Fraassen, B. C. 2008. *Scientific Representation: Paradoxes of Perspective*. Oxford: Oxford University Press.

Van Fraassen, B. C. In Press. 'On Stance and Rationality', *Synthese*. (DOI: 10.1007/s11229-009-9520-1)

Watkins, J. W. N. 1968. 'Non-Inductive Corroboration', in I. Lakatos (ed.), *The Problem of Inductive Logic (Studies in Logic and the Foundations of Mathematics Vol. 51)*. Amsterdam: North-Holland, pp. 61–66.

Watkins, J. W. N. 1969. 'Comprehensively Critical Rationalism', *Philosophy* 44, 57–62.

Watkins, J. W. N. 1971. 'CCR: A Refutation', *Philosophy* 46, 56–61.

Watkins, J. W. N. 1984. *Science and Scepticism*. London: Hutchinson.

Watkins, J. W. N. 1997. 'Popperian Ideas on Progress and Rationality in Science', *The Critical Rationalist* 2. URL: http://www.eeng.dcu.ie/~tkpw/tcr/volume-02/number-02/v02n02.html.

Whewell, W. 1860. *On the Philosophy of Discovery*. London: John W. Parker and Son.

Williamson, J. O. D. 2005. *Bayesian Nets and Causality: Philosophical and Computational Foundations*. Oxford: Oxford University Press.

Williamson, J. O. D. In Press. 'Objective Bayesianism, Bayesian Conditionalisation, and Voluntarism', *Synthese*. (DOI: 10.1007/s11229-009-9515-y)

Williamson, T. 2002. *Knowledge and Its Limits*. Oxford: Oxford University Press.

Williamson, T. 2007. 'How Probable Is an Infinite Sequence of Heads?', *Analysis* 67, 173–180.

Wittgenstein, L. 1963. *Philosophical Investigations*. Translated by G. E. M. Anscombe. Oxford: Blackwell.

Worrall, J. 1988. 'The Value of a Fixed Methodology', *British Journal for the Philosophy of Science* 39, 263–275.

Worrall, J. 1989. 'Structural Realism: The Best of Both Worlds?', *Dialectica* 43, 99–124.

Zahar, E. 1995. 'The Problem of the Empirical Basis', in A. O'Hear (ed.), *Karl Popper: Philosophy and Problems*. Cambridge: Cambridge University Press, pp. 45–74.

Index

For Product Safety Concerns and Information please contact our EU
representative GPSR@taylorandfrancis.com
Taylor & Francis Verlag GmbH, Kaufingerstraße 24, 80331 München, Germany

www.ingramcontent.com/pod-product-compliance
Ingram Content Group UK Ltd.
Pitfield, Milton Keynes, MK11 3LW, UK
UKHW022120190926
475702UK00013B/1545
* 9 7 8 0 4 1 5 8 5 0 0 6 3 *